GENDERED MOBILITIES

I dedicate this book to my grandparents,
Satyaranjan Sahay (1929–1994) and Vimla Sahay
– Tanu

Gendered Mobilities

Edited by

TANU PRIYA UTENG
Norwegian University of Science and Technology, Norway

TIM CRESSWELL
Royal Holloway, University of London, UK

ASHGATE

Published by
Ashgate Publishing Limited
Gower House
Croft Road
Aldershot
Hampshire GU11 3HR
England

Ashgate Publishing Company
Suite 420
101 Cherry Street
Burlington, VT 05401-4405
USA

Ashgate website: http://www.ashgate.com

British Library Cataloguing in Publication Data
Gendered mobilities. - (Transport and society)
1. Spatial behavior 2. Sex differences (Psychology)
3. Transportation - Social aspects 4. Feminist geography
I. Uteng, Tanu Priya II. Cresswell, Tim
304.2'3'082

Library of Congress Cataloging-in-Publication Data
Gendered mobilities / [edited by] Tanu Priya Uteng and Tim Cresswell.
p. cm. -- (Transport and society)
Includes index.
ISBN 978-0-7546-7105-3 (alk. paper)
1. Transportation--Social aspects. 2. Freedom of movement--Sex differences. 3. Travel--Sex differences. 4. Spatial behavior--Sex differences. 5. Wireless communication systems--Social aspects. 6. Social mobility--Sex differences. 7. Movement (Philosophy) I. Uteng, Tanu Priya. II. Cresswell, Tim.

HE151.G339 2008
305.3--dc22

2007034128

ISBN 978-0-7546-7105-3

Printed and bound in Great Britain by MPG Books Ltd, Bodmin, Cornwall.

Contents

List of Figures

List of Tables

List of Contributors

Nadine Cattan, Senior Researcher, National Center for Scientific Research (CNRS), Paris, France

Tim Cresswell, Professor, Department of Geography, Royal Holloway, University of London, UK

Michaela Fay, Research Associate, School of Geography, Politics and Sociology, Newcastle University, UK

Susan Frohlick, Assistant Professor, Department of Anthropology, University of Manitoba, Canada

Clara Greed, Professor of Inclusive Urban Planning, Department of Planning and Architecture, University of the West of England, UK

Randi Hjorthol, Senior Research Leader, Institute of Transport Economics, Oslo, Norway

Tommi Inkinen, Docent and Senior Lecturer, Department of Geography, University of Helsinki, Finland

Anette Jerup Jørgensen, Cand.scient.soc/PhD. Fellow, Institute for Sociology, Social work and Organisation, Faculty of Social Science, Aalborg University, Denmark

David Kronlid, Associate Professor and Post-doc Fellow, Department of Teacher Education, Uppsala University, Sweden; and Department of Archaeology and Religious Studies, Norwegian University of Science and Technology, Norway

Lesley Murray, Research Fellow, Health and Social Policy Research Centre, University of Brighton, UK

Petter Næss, Professor, Department of Development and Planning, Aalborg University, Denmark, with part-time positions at Institute of Transport Economics, Norway and Oslo University College, Norway

Merritt Polk, Program Head, Section of Human Ecology, School of Global Studies, Göteborg University, Sweden

Elisabeth Scheibelhofer, Faculty Member, Institute for Sociology, University of Vienna, Austria

Mimi Sheller, Senior Research fellow, Centre for Mobilities Research and Department of Sociology, Lancaster University, UK; and Visiting Associate Professor, Department of Sociology and Anthropology, Swarthmore College, USA

Sumeeta Srinivasan, Lecturer and Research Fellow, Urban Systems, Harvard University, USA

Sheela Subramanian, SDIP Facilitator, Chester Le Community Coalition, Scarborough, Canada

Nite Tanzarn, Associate, Department of Women and Gender Studies, Makerere University, Kampala, Uganda

Tanu Priya Uteng, Research Fellow, Department of Civil and Transport Engineering, Norwegian University of Science and Technology, Norway

Acknowledgements

This study forms part of the interdisciplinary project 'Technological Spaces of Mobility' hosted by the Norwegian University of Science and Technology, Trondheim, Norway. We would like to thank Professor Tore Sager, coordinator of the Transport Group of this project, for giving us the opportunity and encouragement to realise the idea of *Gendered Mobilities*. A special note of thanks goes to Professor Margaret Grieco whose enthusiasm was both inspirational and instrumental in making *Gendered Mobilities* a reality in a rather short span of time. Finally, this piece would be incomplete without cheering our contributors who have been a very enthusiastic and cooperative group.

Chapter 1

Gendered Mobilities: Towards an Holistic Understanding

Tim Cresswell and Tanu Priya Uteng

Introduction

Mobilities[1] have truly become the hallmark of modern times. But how this hallmark is experienced and represented is far from stable. On the one hand it is positively coded as progress, freedom or modernity itself; on the other hand it brings to mind issues of restricted movement, vigilance and control. Through these dimensions of freedom and control, the understanding of 'mobilities' has offered a cohesive way of viewing the highly globalised/mobilised world we inhabit today. As Lash and Urry (1994, 252) put it, 'modern society is a society on the move'. Similar ideas have been professed in disciplines ranging from philosophy, physics and astronomy to film, photography, architecture and urban planning. However, on our overtly optimistic journey towards progress, we have finally come to terms with the reality of our limits. We have no choice but to pay heed to the threats of climate change and its direct linkage with various aspects of mobility. Consumption, and its connections to mobility, needs to be revisited. Given such contradictory outlooks on the theme, mobility has become a most elusive theoretical, social, technical and political construct. In order to deal with it in a systematic way, it is necessary to revisit the implications of mobility in a holistic manner.

Understanding the ways in which *mobilities* and *gender* intersect is undoubtedly complex given that both concepts are infused with meaning, power and contested understandings. The concept of gender does not operate in a 'binary' form. It is never given but constructed through performative reiteration. The resultant interpretations of gender are also historically, geographically, culturally and politically different, enabling a certain slippage between the different realms in terms of how genders are 'read'. This point is central to an analysis of how mobilities enables/disables/modifies gendered practices. We can use mobility both as an *archive* and *present*

1 Urry (2004, 28) emphasises the need to separate out rather carefully the nature of the five highly interdependent 'mobilities' that form and reform social life, bearing in mind the massive inequalities in structured access to each of these:

1. Corporeal travel of people for work, leisure, family life, pleasure, migration, and escape.
2. Physical movement of *objects* delivered to producers, consumers, and retailers.
3. Imaginative travel elsewhere through images of places and people on television.
4. Virtual travel often in real time on the internet, so transcending geographical and social distance.
5. Communicative travel through person-to-person messages via letters, telephone, fax, and mobile phone.

indictor of discourses, practices, identities, questions, conflicts and contestations to understand its gendered nuances.

The principle aim of this book is to bring the insights of the current 'mobility turn' in the social sciences to bear on these connections between mobilities and gender. Thinking of mobility holistically, we suggest, allows us to see gender on the move from a different and illuminating angle. It brings research and writing formally held apart into conversation through the connecting strand of mobility, which is but one way of theorising the connections between gender and spatiality.

There are many ways in which gender is spatially produced. Perhaps the most commented on is the binary of public and private which has been mapped on to masculine and feminine, man and woman, in clearly delineated ways and been brought into question by any number of feminist theorists. Here gender is defined, at least in part, spatially – through a geographical image. Another key spatial coding for gender, and the one that lies at the heart of this book, is the dialectics of fixity and flow – of place and mobility. By mobility we mean not only geographical movement but also the potential for undertaking movements (motility) as it is lived and experienced – movement and motility plus meaning plus power. Understanding mobility thus means understanding observable physical movement, the meanings that such movements are encoded with, the experience of practicing these movements and the potential for undertaking these movements. Each of these aspects of mobility – movement, meaning, practice and potential – has histories and geographies of gendered difference. Each of these is in some way constructed in a gendered way and each, in turn, contributes to the production, reproduction and contestation of gender itself. How people move (where, how fast, how often etc.) is demonstrably gendered and continues to reproduce gendered power hierarchies. The meanings given to mobility through narrative, discourse and representation have also been clearly differentiated by gender. Similarly, narratives of mobility and immobility play a central role in the constitution of gender as a social and cultural construct. Finally, mobilities are experienced and practiced differently. Acquiring mobility is often analogous to a struggle for acquiring new subjectivity. This reality is in a continuous state of flux, leading to the changing of contours in the relationships between gender, mobilities and shifting subjectivity. Consider just a few of the arenas in which gender and mobilities intersect.

To begin with, we might think of the mechanics of human (and animal) reproduction. Emily Martin considers the language of scientific textbooks describing the human reproductive process. In particular she notes how the mobility of sperm has been described with awe while the relatively stationary egg has been equated with passivity. 'It is remarkable how "femininely" the egg behaves and how "masculinely" the sperm. The egg is seen as large and passive. It does not *move* or *journey*, but passively "is transported," "is swept," or even "drifts" along the fallopian tube' (Martin, 1991, 489). This contrasts with the language used to describe the sperm, which are described as fast, mobile, active and streamlined. Only recently has medical science indicated that the relatively stationary egg might be an active partner in the reproductive process (Martin, 1991). Here, as so often, masculinity is coded as mobile and active while femininity is coded as relatively stationary and passive.

Next consider bodily movements. Iris Marion Young, on noting how male and female students used their bodies while engaged in throwing a ball, or walking with books, concluded that they moved in such a way that they focused in on themselves rather than being focused in an outward way on the world (Young, 1990). Boys, when throwing a ball, would use their whole body to launch the ball towards its target while girls would generally use just the arm. The rest of the body would remain stationary. This inhibited mobility, she argued, meant that girls and women were unable to be the phenomenological 'body-subject' of Maurice Merleau-Ponty (Merleau-Ponty, 1962) but remained a 'body-object'. Or, consider Bourdieu's observations of the different ways in which men and women walked in a Kabyle village in Algeria.

> The man of honour walks at a steady, determined pace. His walk, that of a man who knows where he is going and knows he will get there on time, whatever the obstacles, expresses strength and resolution, as opposed to the hesitant gait (...) announcing indecision, half-hearted promises (...) the fear of commitments and inability to fulfil them. It is a measured pace, contrasting as much with the haste of the man who 'walks with great strides', like a 'dancer', as with the sluggishness of the man who 'trails along'. (Bourdieu, 1990, 70)

Here, as with Young, it is not just a suggestion of women as static and men as mobile that is striking, but the different ways in which mobility is embodied and enacted. Feminine mobilities are different from masculine ones. What's more, this difference acts to reaffirm and reproduce the power relations that produced these differences in the first place.

Similarly, research on gendered travel behaviour patterns have established their spatial variations. Feminist geographers and others have long insisted that analysis of daily travel patterns between home and work cannot be gender blind and that there are very significant differences between the longer distance and more direct daily travel patterns of men in the modern west and the more complicated but shorter distance travel patterns of women who not only go to work but often have to drop children off with schools and childcare as well as dealing with shopping, medical visits and a long list of other, typically feminised, routine events (Law, 1999; Hanson and Pratt, 1995). Studies establish that gender-differentiated roles related to familial maintenance activities place a greater burden on women relative to men in fulfilling these roles resulting in significant differences in *trip purpose, trip distance, transport mode* and *other aspects of travel behaviour* (which includes *different times*, to *different locations* over *different distances*) (Erickson, 1977; Andrews, 1978; Hanson and Hanson, 1981; Howe and O'Connor, 1982; Fagnani, 1983; Fox, 1983; Pas, 1984).

More women than men use public transport, yet this relationship is never explored beyond the domain of transport planning. It has never received attention as an artifact or augmentation as an infrastructure anywhere comparable to that of the car as an object of desire and roads (and other related infrastructure) as a means of catering to freedom. The relationship between men and cars is well known (in the UK there is even a free digital channel called 'Men and Motors'!). Virginia Scharff, in her book *Taking the Wheel*, explored the history of women as drivers, noting the long-standing association between masculinity and driving and the difficulties women faced when taking to the road (Scharff, 1991). Indeed, the early history of the automobile in the United States was marked by attempts to provide alternative and slower means of automobility for women. Foremost amongst these was the electric car. One automotive columnist, C.H. Claudy, believed that the electric car was perfect for women, offering them a safe vehicle with a limited radius, which was perfect for social and domestic tasks.

> What a delight it is to have a machine which she can run herself with no loss of dignity, for making calls, for shopping, for a pleasurable ride, for the paying back of some small social debt. (Quoted in Scharff, 1991, 41)

Compare this polite and slow feminised automobility to the predominantly masculine representations and practices of driving at the time where 'Cars served as private space. Only in private cars could proper middle-class men swear at complete strangers. Men smoked as they pleased in cars. They rapidly became spaces for sexual conquest. Driving, even in traffic, could be made competitive and aggressive, a false bravado' (McShane, 1994).

One of the most significant forms of mobility in the modern world is tourism. The roots of modern western world tourism in the renaissance 'Grand Tour' are well known. This tour was a thoroughly masculine endeavour and elements of that masculinity remain in tourist mobilities that are often conceived of as frontier-like activities of exploration and conquest. The feminist scholar, Cynthia Enloe, has commented on the masculinity of tourism (even when conducted by women). She has also underlined the central role of women in an industry where 75% of workers are underpaid and female and which has long been associated with sex and prostitution.

> It is not simply that ideas about pleasure, travel, escape, bed-making and sexuality have affected women in rich and poor countries. The very structure of international tourism *needs* patriarchy to survive. (Enloe, 1989, 41)

Enloe also notes the long history of women travelers in the pre-tourist age. Privileged travelers like Mary Kingsley who set off from England to Africa in 1892 and traveled throughout the continent, for much of the time without a male escort. Many male explorers felt that their world was being trespassed but Kingsley became extremely popular on the lecture circuit between her travels. While she was breaking gendered codes of mobility she was simultaneously reproducing imperialist codes of mobility as she recounted her travels in the 'dark continent' (Blunt, 1994). Gender and mobility are inextricably linked with other formations of power including class, ethnicity and imperialism.

Consider research. Feminist geographic studies of migration reveal that women's migration decisions and experiences are distinct from men's in that women weigh both reproductive and productive labour demands (Silvey, 2000; Radcliffe, 1991; Chant, 1992; Lawson, 1995). A review of immigration literature reveals that scholars who do not place women's subsistence work at the centre of their analysis construct home/host dualistic arguments that oversimplify women's experiences from transnational mobilities. Kibria (1990) further notes that although some scholars do acknowledge that the migration experience does not seriously challenge patriarchal domination, much of their argument is still framed around the idea that host societies provide greater freedom for women than their home countries. The market/family dichotomy seems to be implicit in the home/host binary framework when scholars argue that, because it provides more paid work opportunities, the host society offers women more freedom than the home country (Lamphere, 1986). Both dichotomies do not fully consider the complexity of women's experiences and cultural differences. Morokvasic (1993) demonstrates the disadvantaged position that immigrant and minority women occupy in the European labour markets. 'While migrants, both male and female, often experience a decline in their occupational status, with migration to Europe, women's position is generally worse than men's. This reflects the restructuring of female-dominated employment sectors, involved as unpaid workers in family businesses, limited employment opportunities due to their legal status as 'dependants', and employers' perceptions of their skills' (Willis and Yeoh, 2000, xiv). Such analysis leave us with questions relating to women's capacity for empowerment as amplified or jeopardised in an era of rapidly intensifying global interdependencies and transnational mobility. We are yet to see a clear exposition of how gender and mobilities intersect to create shifting subjectivity from the perspectives of spatial mobility.

It is also important to bear in mind, however, that these general observations have been constantly contested. Women have always been on the move. Mary Kingsley's explorations in Africa, however complicated by class and imperialism, remind us that women have constantly upset gendered expectations about who moves, how they move and where they move. This is perhaps summed up by a bumper sticker that Cynthia Enloe noticed stuck to the backs of cars around the United States which read: 'Good girls go to heaven, bad girls go everywhere.'

Even this very brief wander through different scales and forms of mobility reveals how gender both constitutes mobility and is constituted by mobility in a myriad of ways. This happens both through the opposition of relative flow and relative fixity, where masculinity is coded as mobile and femininity as static, and through the construction of different kinds of mobility that exist in relation to one another (the tourist and the domestic servant for instance). It should be clear from the different realms of research outlined above that many people, in different disciplines, have something to say about the relationship between gender and mobility. What is less clear, however, is if connections have been made across different instances and scales of movement to consider mobilities more generally. The recent mobility turn, or new mobilities paradigm (Sheller and Urry, 2006), in the social sciences and humanities provides a framework to correct this oversight. While transport studies, or history, or migration theory, for instance, all have something to say about things and people on the move, they have tended to fix on particular kinds of mobility without considering mobilities itself in the wider context. One purpose of this book is to bring these approaches into conversation.

This book builds on a need for an exposition of how theories, social norms, technologies and policies come together to carve out differentiated mobilities. This book, by no means, covers all the diverse perspectives needed to generate a coherent picture of gendered mobilities. However it is an ambition to bring the insights of the mobility turn to bear on the question of the processes of gender production in the mobile world in four distinct ways. The first way it does this is to consider different mobilities in different contexts alongside each other. This allows us to see the interactions of gender and mobility at different scales. The second way is to consider different aspects of mobility. Mobility involves the physical movements, which are observable and representable in maps and models. These are the kinds of mobilities traditionally considered by transport planners and migration modelers. Mobility also involves the meanings associated with movement – the narratives and discourses that make movement make sense culturally. These are the aspects of mobility usually considered by philosophers, literary theorists or academics in cultural studies. And mobility involves practice – the embodied and experienced aspects of moving explored in, for instance, performance studies. The third way is through a combination of styles of research that include theoretical, largely empirical and more applied and policy orientated contributions. There is much to be gained from bringing these approaches into dialogue.

The final way is through the diversity of contributors to this book. They come from a variety of disciplinary backgrounds and consider the interactions of gender and mobilities at many scales, from the body to the globe. They also represent a geographical diversity of scholarship, much of it from beyond the Anglo-American world. There is, perhaps, a tendency to see mobility as a high-tech achievement, a tendency that reflects the production of knowledge in the western, and particularly Anglo-American, context. Looking into mobilities in a wider context is a useful corrective to this.

Dialogical reflections

The book is divided into three parts. The first part delves into theoretical constructs evolving from gendered patterns of mobilities. The chapters presented in this part cover a wide range of issues and highlight the complexity of mobility. Ranging from motherhood and risk mobilities on the one hand to misguided spatial models on the other, this part provides a theoretical palate of diverging points of view. David Kronlid's insights on the relationship between mobility and 'moral agency' from a feminist philosophical standpoint in Chapter 2 compel us to question the unexamined facets of the Capability Approach propounded by the Nobel Laureate, Amartya Sen. David Kronlid argues that we should regard social/spatial/existential mobility as a distinct capability, which will have important consequences for research concerning justice and gender in a number of research areas. Sheela Subramanian frames Chapter 3 around the issues of racialised and gendered mobilities and warns us that any laxity in considering spaces and bodies as themselves constituted through race and gender runs the risk of reproducing the very oppression that we attempt to destabilise when we write about them. She borrows from the writings of Judith Butler and Frantz Fanon to explore the recent works on racialised and gendered mobility by anti-racist feminist Canadian scholar, Sherene Razack. The power and politics of mobilities and how space intersects in shaping them is once again reinforced by the arguments of this chapter. Chapter 4 brings a refreshingly new outlook on the intersection of motherhood, risk and everyday mobilities. Lesley Murray highlights how the mothering culture and risk experience are dependent on mobilities as the mobile society is shaped by gender and risk. She further argues that cultures of mothering are not only determined by ideology but through everyday risk and mobility practices. She captures the hidden nuances of how motherhood and automobility intersect in producing 'good' and 'bad' mothers:

> Within some white middle class cultures the car is a coping strategy, a way of overcoming risk and a symbol of good mothering. For others the only way to avoid risk is to remain relatively immobile, thus complying with local cultures of mothering predicated on risk aversion. There can be little doubt that mothering, along with other aspects of gender, is definitively shaped by mobility and its riskiness and in turn that changing intensities of mobility are determining what it is to be a mother.

Michaela Fay explores in Chapter 5 how mobility is experienced, reflected upon and created in the construction of togetherness and belonging in feminist online networks. Based on her research findings of a cyber-ethnographic analysis of the International Women's University and its participants, she argues that such networks, as well as the body of feminist theory, can be seen as a useful tool in order to understand contemporary mobility, which, especially in the context of the academy, increasingly spans geographical movement as well as the crossing of established theoretical and political boundaries. She underpins how an analysis and further elaboration of the interplay between these two modes can have a fruitful input into understanding the concept of mobility itself. Nadine Cattan in Chapter 6 highlights the importance of considering gender in the construction of spatial concepts and the interpretation of spatial behaviour of populations. She further shows how by conveying a conception of space in movement, fluid, unbounded and unfixed, the concept of mobility deconstructs the classical perception of distance and scale. The conclusions are based on an assessment of spatial theories and case studies of student mobility in Europe and home-workplace mobility in the Paris metropolitan area. In the concluding chapter of this section, Anette Jerup Jørgensen intervenes in the discussion on automobility through a challenging and new perspective. She investigates the gendered distinctions in the context of how drivers relate to one another and to legal regulation in traffic. She concludes that male and female drivers have different approaches to automobility and its risks, to the negotiations of what is morally acceptable behaviour, and also to automobile communication in traffic interaction, which is to a great extent dependent on socially constructed gendered ideas in the society.

How and why are mobilities gendered?

The second part of the book looks into case studies drawn from around the globe to answer the questions of how and why mobilities are gendered. It is well accepted now that the production of some kinds of mobilities (that is automobility) often creates immobilities for others (that is public transport users with limited supply, frequency etc.). We are inundated with facts and data regarding the mobility-poor, socially excluded groups of people. Such reports are often restricted to answering how such groups are mobility-poor but only a few digress to touch the messy question of why they are mobility-poor. Restricting ourselves to looking into the gendered aspect of mobility differentiation, the chapters in this section attempt to answer both 'how' and 'why' questions. In the first chapter of this section, Elisabeth Scheibelhofer looks into how gender still matters and to a great extent dictates the mobility aspirations of European scientists working abroad. She compares the personal relations of mobile women from a developed country with the pitfalls associated with the mobile lives of women from developing countries revealing differences as well as similarities: the family's economic well-being or personal-political freedom is not the most pressing motivation to lead a mobile life, as is frequently the case in transnational mobility from developing countries. Prima facie, mobile scientists from Western European countries are in general less restricted by visa regulations, economic hardship or discriminatory practices. However, the gendered care for social relations still ascribed to women can play a key role when looking at mobilities of women from Western Europe, as her study demonstrates. Chapter 9, titled 'I'm more sexy here…', looks into the ethnographies of travel illuminating the specificities and complexities of how women assert a sexual identity and agency through contemporary touristic mobilities. Susan Frohlick's analysis of the narratives of Western female travellers in Costa Rica highlights nuances of gendered mobilities in which sexual agency and desire give new meanings to the intersection of place, image, norms, moral agency and mobilities. Her concluding sentence entices our view of gendered mobilities 'multiple travel mobilities incite the expression of a "mobile sexuality", where one can feel "more sexy" on one side of a border than the other'! From here on, we follow a string of four chapters dedicated to analysing different facets of interaction between transport and gendered mobility patterns. From the 1960s onwards, research in transportation and planning carved out a new and focused direction under the umbrella of 'gender and transport'. Attention to transport offered a way to link discussions of gender relations, transport systems, public and private spaces, accessibility, and the spatial and temporal organisation of human activity (Law, 1999, 567). However, Law (1999) notes that the field is still largely defined in terms of travel behaviour and policy, eventually stagnated by a relatively limited range of themes (primarily a singular focus on women's typically shorter worker trips). Could this have occurred because of a biased comprehension of mobilities of which transport is just the revealed part?

The understanding that eluded transport planners and geographers equally for a considerable period of time is that the frame within which 'transport' operationalises lies in the broader context of mobilities. Though transport and mobility are very often used in a synonymous manner, they have distinct connotations. Mobility is a contextualised phenomenon whereas transport is just the revealed part of it. The concept of mobility entails the 'potential aspect' thus possessing an inherent knowledge of the potential trips that are/were *not* made due to constraining factor(s) (social, cultural, technological, infrastructural, political and financial). Concurring with the way Law (1999, 568) envisages it, a better way to address 'gender and transport' is through reframing the issues of transport as part of a larger project, namely, analysing the social, cultural, technological, infrastructural, political and financial geographies of mobility. Chapters 10, 11, 12 and 13 have attempted to move in this direction. In Chapter 10, Sumeeta Srinivasan uses the travel diary data for Chengdu, China and Chennai, India to highlight the differences in spatial patterns of accessibility of low-income women and men. Writing against the transport planners' most sought-after solution of constructing new roads or overpasses as a way of lowering overall travel times, she emphasises the need to improve local accessibility through neighbourhood level planning. She reinstates the imperatives drawn from environmental, social and economic perspectives, stating that a primary goal of transport planning should be to improve the accessibility of low-income households and women, who have to walk or bike for both work and non-work activities. Nite Tanzarn, in Chapter 11, explores ways in which gender structures women's and men's mobility patterns in metropolitan Uganda. She further examines how transportation structures and systems create, reproduce and sustain systemic differences in material circumstances between women and men and reinforce women's exclusion and subordination. She gives voice to similar problems afflicting the entire developing world where the knowledge that improving women's mobility and access has the potential to transform the prevailing unequal gender relations is finally gaining ground. The chapter also argues that whereas space is not intrinsically gendered, the inequitable positioning of women in society structures women's and men's utilisation of urban space over time often resulting in exclusion based on gender. In Chapter 12, Petter Næss broaches upon the influences of urban structure on travel. In the widely discussed sustainable city models, this aspect has not been given adequate attention and needs further detailing. Petter Næss highlights that the attempts made in the 1980s to open a gender equality debate within the field of transportation was pioneering, as the 'feminine' urban model emphasised proximity between the different facilities of the city as a strategy where the inhabitants would not have to choose between a high car dependency and constrained opportunities for choice. He concludes that both from a feminist and sustainability perspective, such an urban developmental path should be encouraged. Chapter 13 presents some insights from Norway on the problem which Doreen Massey frames in the following way: 'every time someone uses a car, and thereby increases their own mobility, they reduce both the social rationale and the financial viability of the public transport system – and thereby also potentially reduce the mobility of those who rely on that system' (Massey, 1994, 150). Through touching various related issues ranging from environment and space allocation to distribution of household responsibilities, Randi Hjorthol concludes that the study of men's and women's daily travel patterns can be seen as a 'barometer' of the degree of equality between men and women in society. She adds that even if the number of trips has increased for women during the period of analysis so this is now on the same level as for men, women have a much more limited 'space of action' than men in Norway, which might restrict their choice in the labour market and thus retain the differences.

Seeking grounds for future policies

The third and concluding part of the book attempts to highlight some interdisciplinary ideas that have high relevance for policy making. Policies governing information technology and transport as well as urban planning and design can have a direct and immediate impact on the gendered access to opportunities and consequently carve out sustainable mobilities. In Chapter 14, Tommi Inkinen takes us through the interaction of gender and technology executed through a backdrop of Finnish information society policy. The chapter examines the ways in which the social dimension has been taken into account and its intersection with gender issues in the latest policy documents. His reality check confirms the hypothesis concerning the invisibility of gender in information society strategies. His assessment reveals that the present information society needs a constant involvement with the dimensions of work-life conditions, social equality resulting in safer environments and appreciation of everyday practices, as well as recognition of 'small' things in the creation of policy guidance. From Finland to another Scandinavian country, Sweden, in Chapter 15 Merrit Polk reviews the last five years of 'ineffectual' attempts to mainstream gender in the Swedish transport sector. The chapter gives us a flavour of the pitfalls in the process and the finer details to be borne in mind while undertaking similar policy making efforts. Touching upon the issue of politics and power of mobility, she advises that within transport policy, structural and organisational distributions of power must be analysed and dealt with separately from differences in individual needs and values. She underlines that the goal is to understand the practical repercussions of the masculinity/technology black box on the empirical context of the transport sector and technology as such, and analyses what impact this has for achieving gender equality. The chapter concludes by identifying the gaps in the theoretical and empirical research that is needed to achieve this end. In Chapter 16, Clara Greed revisits the concept of sustainability and emphasises the need to take gender considerations into account in all aspects of urban and transportation planning policies. The chapter weaves its conclusions through taking myriad examples predominantly drawn from the Anglo-American tradition of urban planning. The chapter reinstates the direction shown by Petter Næss in Chapter 12 which points towards the 'city of everyday life', which can be defined as the city of short distances, mixed land uses and multiple centres as the ideal objective that would fully take into account gender considerations. She contends that such a city structure would reduce the need to travel, be more accessible and sustainable, whilst creating higher quality of urban environment for all. In the Epilogue, Mimi Sheller recapitulates mobilities by touching upon its gendered discourses, geographies and technologies. She asserts the importance of gender in the planning, design, and practice of sustainable mobilities. Reverting back to the limits of our consumption, she posits that the solution to contemporary problems of mobility will not be found without sustained attention to women's and men's differential mobilities, to the gendering of design and planning processes, and to the gendered underpinnings of discourses of mobility.

References

Andrews, H.F. (1978), 'Journey to work considerations in labour force participation of married women', *Regional Studies* 12, 11–20.

Blunt, A. (1994), *Travel, Gender and Imperialism: Mary Kingsley and West Africa*. Edited by M. Dear, D. Gregory and N. Thrift, Mappings (New York: Guilford).

Bourdieu, P. (1990), *The Logic of Practice*. Translated by R. Nice (Stanford CA.: Stanford University Press).

Chant S. (ed.) (1992), *Gender and Migration in Developing Countries* (London: Belhaven Press).

Enloe, Cynthia H. (1989), *Bananas, Beaches & Bases: Making Feminist Sense Of International Politics* (London: Pandora).

Erickson, J. (1977), 'An analysis of journey-to-work for women', *Social Problems* 24, 428–35.

Fagnani, J. (1983), 'Women's commuting patterns in the Paris region', Tijdschrift voor Economische en Sociale Geografie 74, 12–24.

Fox, M.B. (1983), 'Working women and travel: the access of women to work and community facilities', *Journal of the American Planning Association* 49, 156–70.

Hanson, S. and Hanson, P. (1981), 'The impact of married women's employment on house-hold travel patterns: a Swedish example", *Transportation* 10, 165–83.

Hanson, S. and Pratt G.J. (1995), *Gender, work, and space, International studies of women and place* (London, New York: Routledge).

Howe, A. and O'Connor, K. (1982), 'Travel to work and labour force participation of men and women in an Australian metropolitan area', *Professional Geographer* 34, 50–64.

Kibria, N. (1990), 'Power patriarchy, and gender conflict in the Vietnamese Community', *Gender & Society* 4, 9–24.

Lamphere, L. (1986), 'Working mothers and family strategies: Portuguese and Colombian women in a New England community', in R.J. Simon and C.B. Brettell (eds), *International migration: The female experience* (Totowa, NJ: Rowman & Allanheld).

Lash, S. and Urry, J. (1994), *The Economies of Signs and Space* (London: Sage Publications).

Law, R. (1999), 'Beyond 'women and transport': Towards new geographies of gender and daily mobility', *Progress in Human Geography* 23 (4), 567–88.

Lawson, V. (1995), 'The politics of difference: Examining the quantitative/qualitative dualism in post-structuralist feminist research', *Professional Geographer* 47, 449–57.

Martin, E. (1991), 'The egg and the sperm: How science has constructed a romance based on stereotypical male-female sex roles', *Signs: Journal of Women in Culture and Society* 16 (3), 485–501.

Massey, D. (1994), *Space, Place, and Gender* (Minneapolis: University of Minnesota Press).

McShane, C. (1994), *Down the asphalt path: the automobile and the American city, Columbia history of urban life* (New York: Columbia University Press).

Merleau-Ponty, M. (1962), *The Phenomenology of Perception*. Translated by C. Smith (London: Routledge and Kegan Paul).

Morokvasic, M. (1993), 'In and out of the labour market: Immigrant and minority women in Europe', New Community 19 (3), 459–83.

Pas, E. (1984), 'The effect of selected socio-demographic characteristics on daily travel behaviour', *Environment and Planning A* 16, 571–81.

Radcliffe, S.A. (1991), 'The role of gender in peasant migration: conceptual issues from the Peruvian Andes', *Review of Radical Political Economics* 23, 129–47.

Scharff, V. (1991), *Taking the wheel: women and the coming of the motor age* (New York: Free Press).

Sheller, M. and Urry, J. (2006), 'The new mobilities paradigm', *Environment and Planning A* 38 (2), 207–26.

Silvey, R.M. (2000), 'Stigmatized spaces: gender and mobility under crisis in South Sulawesi, Indonesia', *Gender, Place and Culture* 7 (2), 143–61.

Urry, J. (2004), 'Connections', *Environment and Planning D* 22, 27–37.

Willis, K. and Yeoh, B. (eds) (2000), *Gender and Migration*, (Camberley: Edward Elgar 'International Studies in Migration' series).

Young, I.M. (1990), *Throwing Like a Girl and Other Essays in Feminist Philosophy and Social Theory* (Bloomington: Indiana University Press).

PART 1
Dialogical Reflections

Chapter 2

Mobility as Capability

David Kronlid

Introduction

The aim of this chapter is to clarify some of the philosophical underpinnings of 'mobility' at the intersection of mobility, social justice, and gender.[1] This is accomplished by discussing *mobility as capability* and by arguing that there are good reasons to regard mobility as a capability.[2]

Health, education and political participation are closely related to social and spatial mobility in terms of well-being and freedom. Being socially and spatially mobile is generally seen as one central aspect of women's well-being.

Taking a departure from capability research, I suggest that the capability approach suffices as a moral philosophical underpinning of what it means to be mobile. Furthermore, I suggest that the significance of mobility for individual well-being is strengthened by a capability approach to mobility.

Firstly, the capability approach and the rapprochement of gender, mobility, and capability research[3] are briefly introduced. This is followed by an analysis of mobility as capability, including an introduction to the ongoing discussion of capability research on which capabilities count. This analysis clarifies the relevance of social and spatial mobility for the capability approach and highlights the importance of a third and overlooked aspect of what it means to be mobile, existential mobility. Finally, I suggest that defining mobility as capability can help to further develop research on mobility, capability and gender and therefore strengthen the claim that mobility is intrinsic to men's and women's well-being.

---

1 I do not analyse the concept of gender in this chapter. Nor do I take a stand on the feminist scholarly debate concerning the meaning of gender, sex, male, female, and related concepts. Rather, I hope that the theoretical suggestions I make are coherent with different theoretical feminist standpoints.

2 Needless to say, there is not enough space here to discuss and clarify all relevant aspects of three large areas of research such as 'mobility', 'gender', and 'capability' or to discuss all or even most of the important implications of such a discussion. Hence, the chapter is an introduction to some significant connections and implications. I am grateful to editors Tanu Priya Uteng and Tim Cresswell, and to Petra Hansson for valuable comments on the text.

3 Research on capability approach is a wide and sprawling interdisciplinary field. See Robeyns (2003) for a thorough theoretical survey of the capability approach, including some of its most central concepts and the differences between Sen's and Nussbaum's approaches.

Gender – mobility – capability

Research that draws attention to gender, mobility and the capability approach already exists. However, research on mobility *as* capability[4] in relation to gender issues is rare.

Before embarking upon the capability approach, it is important to clarify how I use the term 'mobility' in this chapter. My basic understanding of 'mobility' involves the common distinction between potential and revealed movements. Thus, I take as my vantage point an understanding of the meaning of mobility as 'potential movement' (Dunn, 1998; Kaufmann, 2002; Knie, 1997; Sager, 2006; Priya Uteng, 2006).[5]

The capability approach

The capability approach was originally developed by Amartya Sen (1973; 1985; 1993; 2005)[6] as an alternative to utilitarian decision-making frameworks in the fields of development economics and social ethics. Consequently, it is an alternative to measuring 'development' in terms of economic growth and material welfare. Hence, according to the capability approach, the objective of 'development' should be:

> …the promotion and expansion of valuable capabilities (Sen, 1990) and the aim of justice as equality in the space of capabilities. These capabilities are the positive freedom to achieve valuable 'functionings' which range from basic functionings such as being nourished or having shelter to higher level functionings involving friendships, self-respect, and meaningful work. (Alkire and Black, 1997, 263)

Accordingly, the capability approach is a 'broad normative framework for the evaluation and assessment of individual well-being and social arrangements, the design of policies, and proposals about social change in society' (Robeyns, 2003, 94), which focuses on what people are able do and to be. That is, the capability approach focuses on the *ends* of well-being (capabilities) rather than the *means* of well-being (goods):

> Sen argues that our evaluations and policies should focus on what people are able to do and be, on the quality of their life, and on removing obstacles in their lives so that they

---

4 That is, as capability is understood in the capability approach originally developed by Amartya Sen and in the various ways that this approach has been developed by Martha Nussbaum and others.

5 The distinction between potential and revealed movement is analogous to the distinction between capabilities and functions in the capability approach, that is 'capabilities are people's *potential* functionings' (Robeyns, 2003, 63. Emphasis added). Whether or not 'mobility as capability' is theoretically coherent with or corresponds to, for example, Kaufmann's concept of potential movement is a question that needs further discussion.

6 According to Sen (1993, 31), 'functionings represent parts of a state of a person – in particular the various things he or she manages to be and do in life. The *capability* of a person reflects the alternative combinations of functionings the person can achieve, and from which he or she can choose one collection.' Here from Qizilbash (2005, 157).

have more freedom to live the kind of life that, upon reflection, they have reason to value. (Robeyns 2003, 94)

Gender and capability

The capability approach is widely recognised as relevant for studies of gender inequality (Sen, 1995; Nussbaum, 2005a; 2005b; Robeyns, 2003; 2005; Qizilbash, 2005; Agarwal et al., 2003). In this context, one issue under debate is whether Sen's capability approach is too incomplete and vague to 'help us to construct a normative conception of social justice, with critical potential for gender issues', so that we need to 'specify a definite set of capabilities as the most important ones to protect' (Nussbaum, 2003, 33) for measuring gender inequalities. On the other hand, Robeyns argues that although the vagueness or 'underspecified character' (Robeyns, 2003, 67) of Sen's approach makes it 'vulnerable to androcentric interpretations and applications', 'the process by which [a *definite* and universal] list is generated might lack the political [and academic, and perhaps moral] legitimacy needed for policy design' (Robeyns, 2003, 69).[7]

Thus, there is an ongoing academic debate regarding the nature, interpretation and political implementation of the capability approach (Robeyns, 2003).[8] However, these scholars agree that Sen's capability approach avoids the false gender neutrality (Okin, 1989) that other theories of justice are accused of;[9] the capability approach is widely recognised as a suitable approach for analysing gender inequalities.[10]

Gender and mobility

Gender and mobility research is a growing interdisciplinary field involving topics as varied as transport and planning, cultural theory, feminist philosophy, development theory, and postcolonial theory.

---

7 Furthermore, this discussion is related to important questions of adapted preferences in general and gendered preference formation in particular (Nussbaum, 2005, 111–66; Robeyns, 2003, 67).

8 See also Okin (2003) for one feminist philosopher's critical discussion concerning Sen's concept of capability as 'freedom', a critique that Sen responds to in Sen, 2005 (154–5).

9 According to Sen, the capability approach is not a fully-fledged 'theory of justice', because capabilities cannot be the 'sole informational basis' for considerations about process, as one example (Sen, 2005, 156).

10 According to Qizilbash (2005, 151–2), the capability approach 'relates to gender issues at several levels': (1) as a fundamental approach for thinking about gender inequality, (2) as specifically applied to, for example, gender bias in undernutrition and 'measures of gender inequality', which led to the UNDP's Gender Empowerment Measure (GEM) and the 'Gender-related Development Index' (GDI), (3) familial and intrahousehold economic inequality, (4) 'claims of the importance of women's 'agency'', (5) 'women's underevaluation of their own well-being and health, (6) promotion of the cause of gender inequality through a 'dramatic' measure of gender bias – the number of missing women (which relates to mortality difference).

Gender has been recognised in mobility research since the 1970s with varying focus on gendered mobility patterns, such as a focus on work (Law, 1999[11]), immigration (Boyle and Halfacree, 1994; Remennick, 1999), inter- and intra-generational social vertical mobility (Hong Li and Singelmann, 1998), spatial mobility and livelihood strategies (Mandel, 2006), social status quo and residential mobility (Rosenbaum et al., 2002), and space, gender and mobility (Silvey, 2000).

In 1999 Robin Law claimed: '[o]ver the last two decades, a feminist critique of gender-blind transportation research and planning has generated a spate of research into 'women and transport' (567). Law applauds the attention given to gender and its contributions to urban geography, yet she criticises the 'journey-to-work' (570) strand that dominates this research:

> The two decades of scholarship on gender and transport generated a substantial body of research organized around some clearly defined theoretical questions (notably those to do with women's journey to work). Yet an examination of attempts to integrate and summarize the field as a whole shows that the potential of feminist geography of gender and transport has hardly been developed, especially when compared to the outpouring of other research into gender. (Law, 1999, 571)

In the context of geography research, Law suggests that 'future research on the topic must be based on a more systematic treatment of gender as a theoretical concept', which 'identifies aspects of gender as a social category and symbolic code, and links it to aspects of *daily mobility*' instead of aspects of gendered mobility patterns, that is journeys to work (Law, 1999, 567). Law's daily mobility strand involves linking mobility to such aspects as gendered division of labour and activities, gendered access to resources, gendered subject identities, gender as symbolic code, and gendered built environments (Law, 1999, 577–83).

In her feminist and critical analysis of mobility, Janet Wolff argued in 1993 that the metaphors of mobility vocabulary of travel in cultural criticism[12] are gendered, that 'there is an *intrinsic* relationship between [constructed] masculinity and travel' (Wolff, 1993, 230). Wolff's analysis draws on first-person narratives of the women of the Beat generation, of what it was like to be 'off the road' (Wolff, 1993, 228). Furthermore, Cresswell's (1993) geographical reading of mobility as resistance in Jack Kerouac's *On the Road* also highlights gender aspects of travel metaphors and how travel is given meaning through both establishing and resisting cultural gender identities.

11 See Law (2003) for a summary and analysis of gender and mobility research, including references.

12 Cultural criticism includes post-colonial criticism, postmodern theory, and poststructuralist theories of the subject.

Mobility as capability

Recently,[13] the ethical content of mobility has been addressed in terms of capabilities and functionings and the capability approach, the mobility discourse and feminist research have been drawn to one another in a rapprochement of mobility discourse (Priya Uteng, 2006; Cresswell, 2006, 741–2) and feminist political philosophy (Robeyns, 2003). This fusion of gender, mobility and capability highlights and specifies the ethical dimension of mobility in terms of social exclusion and discrimination and raises important theoretical questions concerning the nature of mobility as capability.

Tim Cresswell suggests, following Peak and Ray (2000) that social justice involves 'where people begin – gender, sexuality, race' and where people are in the sense that their location in marginalised communities 'places them at the margins of visibility for justice' and that such 'uneven geographies of oppression are also evident in people's *differential abilities to move*' (Cresswell, 2006, 741–2. Emphasis added). Tanu Priya Uteng suggests that mobility can be seen as a capability and that mobility thus could be studied as one of Amartya Sen's 'realized capabilities', or functionings (Priya Uteng, 2006, 445[14]). Accordingly, in a study of the mobility functionings of non-western immigrant groups in Norway, Priya Uteng concludes:

> We need to focus on enhancing the capabilities of the immigrant group and setting them free to become productive participants in the civil society. Potential mobility options will not only give them a chance to be active agents of development in a democratic society, but will also transform them into assets for their host country. (Priya Uteng, 2006, 460)

In an article on the question of selecting relevant capabilities for studying gender inequalities in the space of capabilities, Ingrid Robeyns suggests that mobility should be one element of a capability list (Robeyns, 2003, 72, 74, 81–2, 85). In addition, mobility, although not always explicitly stated, is implicated in several of the lists presented (Nussbaum, 2003; 2005; Alkire and Black, 1997).

The rapprochement of gender, mobility and the capability approach raises the following questions. First, is it reasonable to regard mobility as a capability in a Senian sense of the term? Answering this question calls for a discussion about theoretical and practical criterion for a capability to be included in a list of capabilities. Second, how can we use mobility as capability in research on gender and mobility in the space of capabilities?

---

13 Note that Sen in his 1979 Tanner lecture 'went into the relevance of 'the ability to move about'…' (Sen, 2005, 158).

14 See also Hong Li and Singelmann (1998) on gender, social mobility, and functionings and Rosenbaum, et al. (2002) on residential mobility and opportunities.

Lists of capabilities

One central question in the capability literature is: what capabilities count?[15] Although Sen is not against capability lists *per se*,[16] his approach is deliberately incomplete (Sen, 2005; Alkire and Black, 1997, 264; Robeyns, 2003, 62). On that ground, Martha Nussbaum[17] (2003, 2005, 178–9), Alkire and Black (1997), Robeyns[18] (2003) and others[19] have presented capability lists in order to operationalise the capability approach.

A list of capabilities indicates what people are able to do and be, and thus indicates what is intrinsically valuable, that is valuable for its own sake (Robeyns, 2003, 61–2) and expresses 'a bare minimum of what respect for human dignity requires' (Nussbaum, 2005, 5).

One argument for capability lists is that the *incompleteness* (Nussbaum, 2005, 13–14) and/or *vagueness* (Nussbaum, 2003, 46–7) of Sen's capability approach lacks 'critical bite when it comes to assessing norms' (Qizilbash, 2005, 159). Hence it is argued that we need a *definite* and universal list of fundamental capabilities (Nussbaum, 2003; 2005; Qizilbash, 2005, 157). However, a definite and universal list is criticised for *over specification* (Alkire and Black, 1997, 268; Robeyns, 2003), which implies that a definite list might not take those in need of such a list into consideration.[20] Here, based on the idea that a definite list of capabilities is reasonable,[21] I will consider whether mobility should be on a list of capabilities, that

15 Another central question concerns whether the capability approach is too individualistic or not (Robeyns, 2003, 105).

16 Please note that Sen is only critical of 'any proposal of a grand mausoleum to one *fixed* and *final* list of capabilities' (Sen, 2005, 160. Emphasis added).

17 Martha Nussbaum's list of capabilities is as follows: (1) Life, (2) Bodily Health, (3) Bodily integrity, (4) Senses, Imagination, and Thought, (5) Emotions, (6) Practical reason, (7) Affiliation, (8) Other species, (9) Play, and (10) Control over one's environment (Nussbaum, 2005, 78–80).

18 (1) Life and physical health. (2) Mental well-being. (3) Bodily integrity and safety. (4) Social relations. (5) Political empowerment. (6) Education and knowledge. (7) Domestic work and non-market care. (8) Paid work and other projects. (9) Shelter and environment. (10) Mobility. (11) Leisure activities. (12) Time-autonomy. (13) Respect. (14) Religion (Robeyns, 2003, 71–2).

19 See also Alkire (2002) and 'Dasgupta, 1993; Desai, 1990; Griffin, 1991; Gough and Thomas, 1994; Qizilbash, 1996; Sen and Anand, 1994; Stewart, 1990' (Alkire and Black, 1997, 269).

20 In other words, that the needs and opinions of those in need of such a list are not taken into consideration when formulating what is supposed to cohere with their meaning of a good and valuable life in dignity.

21 One such theoretical reason is that Nussbaum's *modified universalism* (2005) coheres with Tim Cresswell's (2006, 736–7) description of mobility as both a universal phenomenon and a contextual phenomenon. Furthermore, such a modified universalist approach is coherent with Sen's argument that '[a] general approach can be used in many different ways, depending on the context and the information that is available. It is this combination of foundational analysis and pragmatic use that gives the capability approach its extensive reach', and still keeps its 'critical bite when it comes to assessing norms' (Sen, 1999, here from Qizilbash,

is whether it is reasonable to regard 'mobility' as intrinsically valuable to human flourishing.[22]

Mobility on the list

Mobility can be described as a capability in general terms. Being mobile is a real and genuine opportunity (Priya Uteng, 2006, 445) and a part of what people are able to do and to be. However, in order to show why mobility should be regarded as capability in the Senian sense I will turn to two sequential lines of evaluation suggested by Alkire and Black (1997, 269); *practical reason* and *theoretical reason*.

Practical criteria

According to a principle of practical reason, 'determining the dimensions of human well-being requires people to recognize, on the basis of their own experience, the truthfulness or otherwise of the particular claims' (Alkire and Black, 1997, 269). This line of evaluation includes taking into consideration the fact that abilities and functionings (and the lack thereof) are situated in certain practices[23] located in ecological, economic, and social contexts. Consequently, whether mobility is to be regarded as a capability or not requires an accurate understanding of how local knowledge relates to theoretical assumptions of human well-being (Alkire and Black, 1997, 272).

The practical reasoning is evident in Robeyns' first four of five criteria for the selection of capabilities; *explicit formulation, methodological justification, sensitivity to context, different levels of generality*, and *exhaustion and non reduction* (2003, 70–71).[24]

Following Robeyns, these criteria highlight the fact that when deciding whether mobility should be on a list of capabilities, we should explicate, discuss and defend such a claim both from outside and inside of the scientific community, which includes a need to 'clarify and scrutinize the method that has generated the list and justify this

---

2005, 159; see also Sen, 2005, 160–63 on universalism). Another reason is that Sen's capability approach 'does not stop people from linking the capability approach with some implicit or explicit account of the underlying causes of poverty (for example, unequal land holdings or corruption) and policy prescriptions (for example, land reforms, civil service reforms)' and that it has been implemented by the UN in several programs (Qizilbash, 2005, 159, 158–9). The questions of incompleteness, vagueness, and overspecification are important; however, it is not possible here to analyse them further.

22 Alkire and Black (1997, 268) argue that the term 'functioning' suggests an 'unduly mechanistic account of the human person', and 'prefer the term 'flourishing' – which also communicates the sense that people pursue and participate in but never fully realize' the dimensions of well-being 'once and for all'. Here, I do not take a stand on this matter and use 'functions' and 'flourishing', and 'elements on capability set lists' and 'dimensions of flourishing' interchangeably.

23 See Alkire and Black (1997, 272–4) on connections between simple, complex and incomplete practices and traditions and capabilities.

24 Here, I will only consider Robeyns first four practical criteria.

as appropriate for the issue at hand', for example for gender inequality research (Robeyns, 2003, 70).[25]

In addition, the 'level of abstraction should be appropriate for fulfilling the objectives for which we are seeking to use', in our case, mobility as capability in relation to gender. Furthermore, 'it is important to speak the language of the debate in which we want to get involved', that is the mobility debate. Finally, adding mobility to a list for the purpose of implementable policy proposals, mobility (as the entire list) should be drawn up in two stages; first, as an ideal understanding of mobility 'unconstrained by limitations' of contextual circumstances, and second, one pragmatic list that takes context into account (Robeyns, 2003, 70–72).[26]

In conclusion, there is a strong consensus in capability research that a significant criterion for determining whether a capability should be on the list is to '…involve the affected people and not impose on them a list they simply have to accept, especially when the capability approach is used in political and policy contexts' (Robeyns, 2003, 76, see also Sen, 2005 and Alkire and Black, 1997). Consequently, in relation to gender inequalities and mobility, women and men should be equally involved in determining whether mobility should be on the list.

Theoretical criteria

Based on the idea that the practical criteria are fulfilled[27] I will now look more closely into two theoretical criteria for adding mobility to the list.

The line of evaluation that follows *theoretical reason* involves the following:

> The first is to suggest that a proposed element of the set is not intrinsically valuable but is actually instrumentally valuable – that it is *only* a means to pursuing some other dimension on the list … The second is to suggest that an element of the set is actually part of some other dimension. (Alkire and Black, 1997, 269. Emphasis added.)

Thus we need to clarify if mobility (a) is intrinsically valuable, and (b) if it is part of other dimensions of the list. The latter of these theoretical criteria reflects a list composed of distinct and separate elements; the capabilities are not interchangeable and the need for one cannot be replaced by a larger amount of satisfaction by fulfilling the need for another (Nussbaum, 2005, 81). However, the distinct dimensions are interrelated, for example '[w]omen who can seek employment outside the home have exit options that help them protect bodily integrity from assaults within it' (Nussbaum, 2005, 81). In this sense, the list of capabilities corresponds to human flourishing as a coherent whole.

---

25 Robeyn's line of reasoning is coherent with Sen's idea that a list of capabilities should be constructed out of certain practical and concrete purposes, and should be tested against public reasoning (Sen, 2005).

26 Thus, here I will discuss mobility as capability in this ideal sense. It is important to note that an 'ideal' list of capabilities should be contextually debated, scrutinised, and justified.

27 It does not fall within the aim of this chapter to engage in the kind of empirical research that the line of practical reasoning suggests.

Explicit and implicit mobility in existing lists of capabilities

Researchers in the fields of mobility, feminism, and philosophy are already explicitly and implicitly adding different dimensions of mobility to capability lists, but does this imply that mobility should be regarded as intrinsic to people's well-being?

Robeyns (2003, 81–2) adds mobility as capability number ten on her list of fourteen capabilities.[28] She distinguishes between the capability to move and revealed movement for example 'public transport', 'movement between geographical locations', and 'travelling with pushchairs'). Furthermore, Nussbaum includes '[b]eing able to move freely from place to place' in capability number three on her list, *Bodily Integrity* (Nussbaum, 2005, 78).

However, these lists contain only spatial mobility and no explicit reasons are given as to why other dimensions of mobility are excluded. This can of course have various explanations. One reason might be firm barriers between feminist philosophy and mobility research, which has resulted in a lack of knowledge of, and interdisciplinary work on, the implications of the dissolution of the barrier between social and spatial mobility (for example, Priya Uteng, 2006;[29] Cresswell, 2001) for research on gender inequality and mobility. Another reason might be that the capability of being socially mobile is being treated as a part of other dimensions on these capability lists, which is suggested by the following analysis of the list by Martha Nussbaum (2005).[30]

Both social and existential (to be clarified below) mobility are implicated in Nussbaum's list. At the core of individual social mobility lies the ability to encounter, connect, and engage in relationships with other agents, objects, and places.[31] Hence, to be able to move socially means to have the ability to move (horizontally as well as vertically) in social space (Sorokin, 1959). Following this, social mobility is

---

28 See n. 18 for a summarised version of Robeyns' list of capabilities.

29 See Priya Uteng (2006, 439) for further references on research on the issue of the dissolution of barriers between geographical/spatial/physical movement (geography, urban planning, and transport) and social sinking and climbing (sociology).

30 A third reason could be that perhaps spatial and social mobility operate in the realm of the taken for granted in the space of capabilities. When only spatial mobility is on the list (as in Robeyns' case) or mobility as such is only implicated (as in the cases of other lists), the significance of social mobility is taken for granted in these theoretical approaches. Thus, this may be analogous to how Cresswell (2001, 742) argues that: 'In the United States the right to mobility has operated in the realm of the taken for granted.' Regarding additional capability lists, see Robeyns (2003, 74) in which she presents two additional lists of capabilities alongside her own and Martha Nussbaum's list. *(A) Swedish approach (1987)*: (1) Mortality, (2) Physical and mental health and healthcare use, (3) Employment and working hours, (4) Working conditions, (5) Economic resources, (6) Educational resources, (7) Housing conditions, (8) Political resources, (9) Family and social integration, (10) Leisure and recreation; (B) *Alkire and Black (1997)*: (1) Life, (2) Knowledge and appreciation of beauty, (3) Work and play, (4) Friendship, (5) Self-integration, (6) Coherent self-determination, (7) Transcendence, (8) Other species.

31 Although there is an important difference between social mobility and social interaction, it is hard to imagine the one without the other.

implicated in Nussbaum's capability number two, *Bodily Health*. Here, reproductive health is included, which means that 'people are able to have a satisfying and safe sex life and that they have the capability to reproduce and the freedom to decide if, when, and how often to do so' (Nussbaum 2005, 78, n. 83). Accordingly, being able to have reproductive health means being able to encounter and connect with others and to engage in social, that is sexual and/or reproductive relationships with whatever partner one chooses; thus to be able to engage in social and embodied movements. Such social interaction is further accentuated in capability number three, *Bodily Integrity* (Nussbaum, 2005, 78).

In addition, capabilities number seven, *Affiliation*, and number eight, *Other Species*, further emphasise social embodied interaction, which implies social mobility as capability (Nussbaum, 2005, 79–80).[32]

Both *Affiliation* and *Other Species* include spatial terminology, such as 'live *with* and *toward* others' (emphasis added) and be able to *enter* 'into meaningful relationships'. Thus, this underlines the fact that the ability to be socially (or rather ecosocially) mobile is intrinsic to human well-being.

A further important aspect of social mobility involves basic democratic principles concerning the right to enter (directly or by representation) into the social settings and institutions that execute political choices that affect one's life. This dimension of social vertical or horizontal mobility is highlighted in the political aspect of capability number ten, *Control Over One's Environment*, '[b]eing able to participate effectively in political choices that govern one's life…' (Nussbaum, 2005, 80).

To summarise, this brief analysis shows that Nussbaum's list involves individual social interaction. Further, in so far as these interactions – capabilities of bodily integrity, bodily health, affiliation, control over one's environment etc. – involve horizontal and/or vertical transitions of the individual from one social position to another, social mobility is implicated. This might suggest that social mobility should not be a distinct dimension, because it is obviously a part of other dimensions of human flourishing. However, spatial mobility *is* a distinct dimension on several lists. And mobility research shows that spatial mobility is intimately connected to social mobility so that 'a change in geographical/spatial mobility patterns affects the individual space of options and action, thus producing varying terrains of social mobility' (Priya Uteng, 2006, 439).

This indicates that instead of only keeping spatial mobility on the list because other dimensions of flourishing cover social mobility without considering the multidimensional nature of mobility, we have reason to separate social mobility from these other dimensions and to expand the meaning of mobility as a distinct capability. This argument is supported by the following introduction to the existential dimension of mobility.

32 See for example: 'Being able to … engage in various forms of social interaction […] being able to work as a human being … and entering into meaningful relationships of mutual recognition with other workers' and [b]eing able to live with concern for and in relation to animals, plants, and the world of nature' (Nussbaum 2005, 79–80).

Existential mobility I suggest that mobility should be included as a distinct capability on a list of capabilities, because the capability to be mobile also includes what I refer to as existential mobility.[33]

It is convincingly argued that mobility *is* a multidimensional concept and phenomenon; there are several dimensions of spatial mobility (Urry, 2000), spatial, temporal and contextual dimensions of mobility interact (Kakihara and Sörensen, 2001), and the barriers between spatial and social mobility are still dissolving (Priya Uteng, 2006). Further, it is sometimes argued that for every mode of mobility or dimensions of mobility there are (overlapping) spaces (Sager, 2006, 467). Thus, geographical mobility is situated in geographical space, social mobility in social space, symbolic mobility in symbolic space and so forth. Can this mean that there is potential movement for every space and that if there is existential space there is also existential mobility?

Peter Nynäs suggests that '[t]he concept 'existential place' describes a transformative potential in the relationship between place and the human self that is significant for the understanding of mobility' (Nynäs, in press).[34] Furthermore, existential space is recognised 'as an intermediate area between the individual and the environment … as a potential or transitional space' (Nynäs, in press), which 'is also characterised by the human capacity to nurture and develop the fundamental relatedness between self and other through play, fantasy, creativity and growth (Winnicott, 1971)' (Nynäs, in press).[35]

If there is such a space – existential space – is there also existential mobility? Such a term might apply to abilities to move our mind, heart, and soul; the capabilities and functions needed to make sense of and value the world as we pass through it and it passes through and around us.[36]

---

33 Tim Cresswell (2001, 738) has a point when arguing that philosophical and other focus on the 'flux' and 'flow' of the world have 'sometimes tended to overstate the case for a wonderful new world of mobility'. And obviously we need to be careful not to make mobility into another black box; applied to everything and therefore accomplishing little. See also Law (1999, 573–4) for a feminist critique of mobility metaphors 'glossing over gendered meaning of mobility in western experience' (573). Mobility is also a theme in 'post modern' ethics, where mobility functions signify a shift in the quality of moral outlooks (Baumann, 1993; 1995). Spatial movements as revealed in mobility and space is also a theme in anthropological research, see for example the ground-breaking work of Tim Ingold (2002).

34 Nynäs develops the Finnish architect Juhani Pallasmaa's (2001) conception of existential space.

35 This view is somewhat coherent with the idea of poststructural or discursive subjects: 'The nomadic subject is amoeba-like, *struggling to win some space for itself in its local context*' (Grossberg, 1987, 39). Here from Wolff (1993, 227, emphasis added). What I am suggesting is simply that such struggles include and are conditioned by social/spatial/existential mobility. (Although I firmly believe that ideas of 'the fluid and the provisional nature of the subject' (Wolff, 1993, 227) should include a theory of how provisional boundaries between self and other are overcome, fading, and rebuilt – always rebuilt.)

36 See Kronlid (in press) for a more detailed exploration of the concept of existential mobility and meaning making processes.

I suggest that this is a path worth exploring if we want to strengthen the case for mobility as a distinct element or dimension of the capability approach.

'Existential space' might be regarded as one dimension of mobility, and I suggest that to be able to be existentially mobile applies to what we usually refer to as 'inner' capabilities – emotions, concepts, and imaginations – in order to create and sustain a worthy and meaningful existence.[37]

Individual meaning-making processes are to a large extent about having the freedom and opportunity to imagine potential futures for yourself and to implement these in the present. Such processes are about being emotionally, mentally and religiously free and are implicated in Martha Nussbaum's list concerning Senses, Imagination and Thought, Emotions, Play and Practical Reason. Here we find abilities to 'imagine' in connection with 'experiencing and producing works and events of one's own choice, religious, literary, musical, and so forth'; to use your mind 'with respect to both political and artistic speech, and freedom of religious exercise'; to 'have attachments to things and people', 'to love, to grieve, to experience longing, gratitude and justified anger'; to be able to 'form a conception of the good and to engage in critical reflection about the planning of one's life'.

These examples suggest that people are capable of moving between the temporary barriers of the space of the 'inner self' and social and geographical space. Potential and revealed movements create the intermediate space between self and environment (existential space), which is needed for the individual to imagine and to dread potential identities and future lives. Thus, existential mobility as capability seems to be intrinsic to knowing and acting upon what we 'are able to be and to do' (Robeyns, 2003, 62).[38]

To conclude, I suggest that the fact that mobility is already being acknowledged as capability in capability and mobility research suggests that there are reasons to regard mobility as intrinsic to well-being.[39] However, research on the multidimensional nature of mobility and on existential space, as well as the fact that social and existential mobility are implicated on several lists, suggests that mobility in *all* its dimensions should be regarded as a distinct dimension of human flourishing.[40] Please note that the idea of mobility as capability does not mean that all kinds of *revealed*

---

37 This idea of existential mobility is inspired by research on cognitive and perceptive processes situated in our bodies' movements in and through landscapes (Overgaard, 2005) argued in ecological psychology, and by the deep intuitive notion and ecological fact that '[m]obility has the status of a fact of life. To be human, indeed, to be animal, is to have some kind of capacity for mobility. We experience the world as we move through it (Cresswell, 2006, 737)' (Ingold, 2002; Kronlid, 2006).

38 See also Robeyns, 2003, 96 for a discussion regarding the fact that the capability approach 'covers *all* dimensions of human well-being' (emphasis added).

39 This argument rests on the premise that the reasons given in the capability research in question are acceptable reasons; this is an important question of epistemological justification that cannot be dealt with here.

40 An additional reason for accepting mobility as capability is that there seems to be a conceptual and normative affiliation between the conceptual frameworks of capability and mobility research. Such affiliations should not be underestimated from the perspective that the rapprochement of mobility and gender is a growing interdisciplinary field.

movements are intrinsic to well-being, but that mobility as potential social/spatial/existential movements are.

Evaluating gender inequalities in the space of mobility capabilities

Based on the idea that it is reasonable to regard mobility as a capability, I suggest the following research areas at the crossroads of mobility-gender-capability.

Function studies

One obvious outcome is the evaluation of gender inequalities related to distinct and interrelated dimensions of mobility in the space of capabilities (Qizilbash, 2005, 159). Such studies can effectively be conducted in the manner of Tanu Priya Uteng's empirical research on ethnic inequalities, that is as combined quantitative and qualitative analyses of mobility *functionings* (Priya Uteng, 2006). Furthermore, such studies can be combined with critical analyses of structural and particular social, economic and symbolic causes for capability inequalities, including analyses of gendered mobility, such as androcentric vocabulary, in religion, media, fiction, and research.

Responsibility studies

Cause and effect analyses suggest philosophical and empirical studies of moral responsibility. Martha Nussbaum argues (2005) that it is the state's responsibility to provide the social base needed in order for every individual to develop his or her capabilities. However, as Robeyns (2003) suggests, this is not a clear-cut issue. In the context of a growing number of powerful actors (for example media, multinational, global and local corporations, NGOs, IGOs, education) who can influence mobility, the question of who can be held accountable for people's lack of mobility is linked to ideas of institutional and individual duties and rights and cannot be taken for granted; thus, further critical gender research on this matter is needed.

Interdependence studies

The capability approach functions as vantage point for analyses that ask if 'realization of one freedom can constrain realization of another' (Qizilbash, 2005, 154). This is related to the interdependence aspect of capabilities, that is that '[g]aining the freedom to do the things that we have reason to value is rarely something we can accomplish as individuals' (Evans, 2002, 56).

Thus, empirical comparative gender capability analyses can clarify when and why men and women who gain certain mobility capabilities are dependent upon gaining other capabilities. Furthermore, when and why men and women gaining a specific capability constrains other men's and women's capability to be socially, spatially, and or existentially mobile.[41]

For example, Cresswell (2006, 738–9) points out that often the increased revealed mobility of one is built upon the localness of others, as in the cases of the 'global elite' and 'the labour forces who experience travel' (Kaplan, 2002) or of the Philippine workers in Canada (Pratt, 1999).

Interdependence studies include research on the interconnections between social/spatial/existential mobility, and how these interconnections relate to how these dimensions of mobility are gendered in given contexts.

Social exclusion studies

Incorporating the capability approach into mobility and gender research can further develop ongoing descriptive and critical feminist and other research on social exclusion and mobility. Whereas Lyon argues that qualitative and quantitative accessibility-based measurements 'associated to given locations' are needed (as complement to trip-based and activity-based measurements), *access* is basically a dimension of revealed mobility (Lyons, 2003, 340). From a capability approach point of view, accessibility-based analyses should include analyses of connections between access to travel and capability to travel in the same spirit as mobility functionings are analysed as mobility capabilities in action.

Development and environmental studies

One obvious general disadvantage of the capability approach and of mobility research is that it is firmly rooted in development ethics and the realm of the human, which means there is a risk of turning a blind eye to the interconnections between social exclusion and the environmental crisis. Further, ecological dimensions of contexts are not included in the contextual mobility analyses suggested by for example Nijkamp et al. (1990) in Priya Uteng (2006).

Ecofeminists (for example Mies and Shiva, 1993; Warren, 2000; Plumwood, 1997) and social ecologists (for example Zimmerman, 1993) repeatedly argue that there are empirical, semantic, ethical, epistemological, and political connections between the exploitation of humans – and more specifically of women – and the exploitation of nature. Such connections are not sufficiently analysed in either capability or mobility research. Thus, given the fact that revealed motorised mobility is closely related to the present ecological global crisis and its central place in modernity, *and* that the environmental crisis is equally a cultural crisis, the rapprochement of gender, mobility and capability could serve as a vantage point for such theoretical development.

41 Such comparative analyses can be conducted on both the individual and social group level.

In other words, if there are reasons to believe that (and we do have reason to believe this[42]) realising freedom for male citizens means constraining capabilities for female citizens, *and* for nonhuman nature, that *andro*centric mobility and *anthropo*centric mobility are intertwined, then the capability as mobility approach should address this issue and thus clarify the nature and content of such interconnections in the space of capabilities.[43]

Resistance studies

Tim Cresswell (1993) argues that mobilities involve both resistance against and the embrace of existing norms and values in a given context, which a capability approach ought to consider, because any contextual analysis should involve a critical approach to mainstream value systems of the contexts in question.

Priya Uteng (2006, 449–51) argues in favour of equal access to driving licences among citizens in Norway. However, although I agree that being able to own and drive a car in Norway is significant for any number of important activities, such as entering the labour market, to evaluate gendered mobility inequalities in the space of capabilities should also include critical analyses from the perspective of economic, social, *and* ecological dimensions of the context. In other words, any analysis should clarify if parts of the mobility capabilities in question correspond to notions of social climbing and inclusion that are a part of a value system, which is a part of a notion of a non-sustainable conception of 'development'.[44] Hence, once one has the capability to choose whether to have a drivers licence or not, people may actually choose not to, as a way of demonstrating resistance towards 'development strategies [that] are based on the explicit or implicit assumption that the model of 'the good life' is that prevailing in the affluent societies of the North: the USA, Europe and Japan' (Mies, 1993, 55).

Conclusion

In this chapter I have approached gender and mobility from the perspective of the capability approach. I have suggested that the question of whether mobility should be regarded as a distinct capability is a matter of theoretical and practical reasoning and have offered several theoretical reasons for why it is reasonable to add mobility to a list of capabilities.[45] In other words, I suggest that mobility should be regarded as intrinsic to human well-being and that spatial/social/existential mobility should, although interrelated to other capabilities on a given list, be separated from them.

---

42 See for example, Cuomo (1998), Kronlid (2003), McFague (1997), Merchant (1990, 1992, 1996), Warren (2000), and Plumwood (1997).

43 For a discussion of this research, see Kronlid (2006) in *Climate Capabilities*.

44 Here, sustainability refers to ecological, economic, and social sustainability, as well as to the fact that these three dimensions of sustainability are interconnected.

45 I have not discussed whether lists are needed or if capabilities are universal. Rather I have taken at least what Robeyns refers to as ideal lists as the vantage point for my discussion.

One reason for this is that spatial mobility has already been presented as a distinct capability, and social and existential mobility are already implicated in various capability lists. This, taken together with the multidimensional nature of mobility presented in various aspects of mobility research, is perhaps the strongest argument for why mobility, in all its aspects, should be considered a fundamental distinct capability for a life of dignity; to be mobile is intrinsic to human well-being. A second reason is that from an interdisciplinary perspective, an overlooked dimension of mobility seems to be intrinsic to what people are able to be and do; that is, existential mobility (the capability of being emotionally, mentally, and intellectually moveable). I suggest that what I refer to as existential mobility is suggested in several of the capability lists and that because being spatially mobile always also means being socially and existentially mobile, the case for mobility as a distinct capability is strengthened. Finally I have concluded that if it is reasonable to regard social/spatial/existential mobility as a distinct capability, this will have important consequences for research concerning justice and gender in a number of suggested research areas.

References

Agarwal, B. et al. (2003), 'Amartya Sen's Work and Ideas: A Gender Perspective', the special double issue of *Feminist Economics* 9 (2/3) July.

Aldridge, S. (2006) 'Social Mobility: A Discussion Paper', *Prime Minister's Strategy Unit* [website], <http://www.strategy.gov.uk/work_areas/social_mobility/index.asp> (accessed 23 November 2006).

Alkire, S. (2002), *Valuing Freedoms: Sen's Capability Approach and Poverty Reduction* (Oxford: Oxford University Press).

Alkire, S. and Black R. (1997), 'A Practical Reasoning Theory of Development Ethics: Furthering the Capabilities Approach', *Journal of International Development* 9 (2), 263–9.

Attfield, Robin (1999), *The Ethics of the Global Environment*, Edinburgh Studies in World Ethics (Edinburgh: Edinburgh University Press).

Baumann, Z. (1993), *Postmodern Ethics* (London: Blackwell Publishing).

Baumann, Z. (1995), *Life in Fragments. Essays in Postmodern Morality* (London: Blackwell Publishing).

Beck, U. (2001), *Vad innebär globaliseringen. Missuppfattningar och möjliga politiska svar* (Göteborg: Daidalos AB).

Bergmann, S., Hoff, T. and Sager, T. (eds) (in press), *Spaces of Mobility: Essays on the Planning, Ethics, Engineering and Religion of Human Motion* (London: Equinox Publishing).

Boyle, P.J. and Halfacree, K.H. (1994), 'Service Class Migration in England and Wales, 1980-1981: Identifying Gender-Specific Mobility Patterns', *Regional Studies* 29 (1), 43–57.

Cuomo, C. (1998), *Feminism and Ecological Communities. An Ethics of Flourishing* (London and New York: Routledge).

Cresswell, T. (1993), 'Mobility as Resistance: A Geographical Reader of Kerouac's 'On the Road', *Transactions of the Institute of British Geographers*, New Series 18, (2), 249–62.

Cresswell, T. (2001) 'The Production of Mobilities', *New Formations* 43, 4–28.

Cresswell, T. (2006), 'The Right to Mobility: The Production of Mobility in the Courtroom', *Antipode* (Oxford: Blackwell Publishing), 735–54.

Dower, Nigel (1998), *World Ethics. The New Agenda*, Edinburgh Studies in World Ethics (Edinburgh: Edinburgh University Press).

Dunn, J.A. (1998), *Driving forces. The automobile, its enemies, and the politics of mobility* (Washington DC: Brookings Institution Press).

Evans, P. (2002), 'Collective Capabilities, Culture, and Amartya Sėn's Development as Freedom', *Studies in Comparative International Development*, Summer 37 (2), 54–60.

Gough, I. and Thomas, T. (1994), 'Need Satisfaction and Welfare Outcomes: Theory and Explanations', *Social Policy & Administration* 28 (1).

Hong Li, J. and Singelmann, J. (1998), 'Gendered Social Mobility', *Scandinavian Sociological Association, Acta Sociologica* 41, 315–33.

Ingold, T. (2002) *The Perception of the Environment. Essays in livelihood, dwelling and skill*, second printing (London and New York: Routledge).

Kakihara, M and Sörensen, C. (2001), 'Expanding the 'Mobility' Concept', *SIGGROUP Bulletin*, December 22 (3), 33–7.

Kaufmann, V. (2002), *Re-thinking Mobility* (Aldershot: Ashgate).

Knie, A. (1997), 'Eigenzeit und Eigenraum: Zur Dialektik von Mobilitat und Verkehr', *Soziale Welt* 47 (1), 39–54.

Kronlid, D. (2003), *Ecofeminism and Environmental Ethics. An Analysis of Ecofeminist Ethical Theory*, *Uppsala Studies in Social Ethics* 28 (Stockholm: Uppsala University Library).

Kronlid, D. (in press), 'What Modes of Moving Do to Me – Reflections about Technogenic Processes of Identification', in S, Bergmann, T. Hoff, and T. Sager (eds) (in press), *Spaces of Mobility. Essays on the Planning, Ethics, Engineering and Religion of Human Motion* (London: Equinox Publishing).

Kronlid, D. (2006), 'Climate Capabilities. An Ethical Analysis of Anthropogenic Climate Change', Research project application, The Swedish Research Council Formas.

Langan, C. (2001), 'Mobility disability', *Public Culture* 13 (3), 459–84.

Latour, B. (1999) *We Have Never Been Modern* (Cambridge, MA: Harvard University Press).

Law, R. (1999), 'Beyond 'Women and Transport': Towards New Geographies of Gender and Daily Mobility', *Progress in Human Geography* 23 (4), 567–88.

Lyons, G. (2003), 'The Introduction of Social Exclusion into the Field of Travel Behaviour', *Transport Policy* 10, 339–42.

Mandel, J.L. (2006), 'Creating Profitable Livelihoods: Mobility as a 'Practical' and 'Strategic' Gender Need in Porto Novo, Benin', *Tijdschrift voor Economische en Sociale Geografie* 97 (4), 343–63.

McFague, S. (1997), *Super, Natural, Christians. How We Should Love Nature* (Philadelphia: Fortress Press).

Merchant, C. (1980/1990), *The Death of Nature. Women, Ecology, and the Scientific Revolution* (San Francisco: Harper).
Merchant, C. (1992), *Radical Ecology. The Search for a Liveable World* (New York: Routledge).
Merchant, C. (1996), *Earthcare. Women and the Environment* (New York: Routledge).
Mies, M. (1993), 'The Myth of Catching up Development', in M. Mies, and V. Shiva, *Ecofeminism* (London/New Jersey: Zed Books) 55–70.
Mies, M. and Shiva V. (1993), *Ecofeminism* (London/New Jersey: Zed Books).
Nijkamp, P., Reichman, S. and Wegener, M. (eds) (1990), *Euromobile: Transport, Communications and Mobility in Europe. A Cross-national Overview* (Aldershot: Avebury).
Nussbaum, M. (2003), 'Capabilities as Fundamental Entitlements: Sen and Social Justice', *Feminist Economics* 9 (2–3), 33–59.
Nussbaum, M. (2005a), *Women and Human Development: The Capabilities Approach*, 8th printing (Cambridge: Cambridge University Press).
Nussbaum, M. (2005b), 'Women's Bodies: Violence, Security, Capabilities', *Journal of Human Development* 6 (2), July, 168–83.
Nussbaum, M. and Glover, J. (eds) (1995), *Women, Culture, and Development* (Oxford: Clarendon Press).
Nynäs, P. (in press) 'Global Vagabonds, Place and the Self: Psychological affects on work-related mobility on morality and world view', in S. Bergmann, T. Hoff and T. Sager (eds) (in press), *Spaces of Mobility. Essays on the Planning, Ethics, Engineering and Religion of Human Motion* (London: Equinox Publishing).
Okin, S. (1989) *Justice, Gender, and the Family* (New York: Basic Books).
Okin, S. (2003) 'Poverty, Well-being and Gender: What Counts, Who's Heard?', *Philosophy and Public Affairs* 31, 280–316.
Overgaard, K. (2005) 'Fitting Technical Environments to Embodied Opportunistic Minds: Creating Liveable and Safe Environments for Humans', Paper presented at the Embodiment and Environment Conference the 5–8 July 2005 (Oxford: Oxford Brookes University).
Palasmaa, J. (2001), *The Architecture of Image: Existential Space in Cinema* (Helsinki: Rakennustieto).
Plumwood, Val (1993/1997), *Feminism and the Mastery of Nature* (London/New York: Routledge).
Priya Uteng, T. (2006), 'Mobility: Discourses from the Nonwestern Immigrant Groups in Norway', *Mobilities* 1 (3), November, 435–62.
Qizilbash, M. (1996). 'Capabilities, Well-Being and Human Development A Survey', *Journal of Development Studies* 33 (2), 143–62.
Qizilbash, M. (2005), 'Dialogue: Sen on Freedom and Gender Justice', *Feminist Economics* 11 (3), November, 151–66.
Remennick, L. (1999) 'Women with a Russian Accent', in Israel, 'On the Gender Aspects of Immigration', *The European Journal of Women's Studies* 6, 441–61.
Robeyns, I. (2003), 'Sen's Capability Approach and Gender Inequality: Selecting Relevant Capabilities', *Feminist Economics* 9 (2/3), 61–92.

Rosenbaum, J., Reynolds, L. and Deluca, D. (2002), 'How Do Places Matter? The Geography of Opportunity, Self-efficacy and a Look Inside the Black Box of Residential Mobility', *Housing Studies* 17 (1), 71–82.

Sager, T. (2005), 'Footloose and Forecast-Free: Hypermobility and the Planning of Society', *European Journal of Spatial Development*, September, 17, [website] <http://www.nordregio.se/EJSD/refereed17.pdf> (accessed 23 November 2006).

Sager, T. (2006), 'Freedom as Mobility: Implications of the Distinction between Actual and Potential Travelling', *Mobilities* 1 (3), November, 465–88.

Sen, A. (1973), *On Economic Inequality*, reprinted in 1997 (Oxford: Clarendon Press).

Sen, A. (1985), *Commodities and Capabilities*, reprinted in 1999 (Delhi: Oxford University Press).

Sen, A. (1993), 'Capability and Well-Being', in Nussbaum, M. and Sen A. (eds), *The Quality of Life* (Oxford: Oxford University Press), 30–55.

Sen, A. (1995), 'Gender Inequality and Theories of Justice', in M. Nussbaum and J. Glover (eds) *Women, Culture, and Development* (Oxford: Clarendon Press), 259–273.

Sen, A. (2005), 'Human Rights and Capabilities', *Journal of Human Development* 6 (2), July, 151–66.

Silvey, R.M. (2000), 'Stigmatized Spaces: Gender and Mobility under Crisis in South Sulawesi, Indonesia', *Gender, Place and Culture* 7 (2), 143–61.

Sorokin, P. (1959), *Social and Cultural Mobility* (New York: The Free Press).

Stewart, F. (1990), 'Basic Needs Strategies, Human Rights and the Right to Development', *QEH Working Paper Luca D'Angliana*.

Urry, J. (2000), *Sociology beyond Societies: Mobilities for the Twenty-First Century* (London: Routledge).

Warren, K.J. (2000), *Ecofeminist Philosophy: A Western Perspective on What it is and Why it Matters* (Oxford: Rowman and Littlefield Publishers).

Wolff, J. (1993), 'On the Road Again: Metaphors of Travel in Cultural Criticism', *Cultural Studies* 7 (2), 224–39).

Zimmerman, M.E (ed.) (1993), *Environmental Philosophy: From Animal Rights to Radical Ecology* (New Jersey: Prentice Hall, Englewood Cliffs).

Chapter 3

Embodying the Space Between: Unmapping Writing about Racialised and Gendered Mobilities[1]

Sheela Subramanian

Introduction

In July 2005, a woman's body was found on the northern edges of the Canadian city of Edmonton during the search for a missing woman, Liana White. The announcement of this discovery resulted in an interesting politics of gender, 'race' and space, marking a period of time during which women's bodies, mostly of Aboriginal women working in the sex trade, were found in rural areas just outside the city. Following the announcement, there was much media speculation as to whether this body was indeed White's, or a prostitute's, or whether White had perhaps fallen victim to the sex trade serial killer (Harding, 2005, A1; Simons, 2005, B1). After White's husband was charged with the murder, local journalist Paula Simons of the *Edmonton Journal* pointed out what she saw as the absurdity of this last suggestion: 'Liana White was a pregnant suburban mom, a medical clerk at the Royal Alexandra Hospital neo-natal unit, a most improbable victim for a serial killer of prostitutes' (Simons, 2005, B1).

Is there a manner in which some bodies belong to certain spaces and not to others? Paula Simons' quote suggests that not only is there such a relationship of bodies to space, but that there are horrific implications of this belonging. Her words prompt a number of additional questions about bodies and spaces that are explored in this chapter: Is this relationship of belonging related to meanings of 'race' and gender? What role does Simons' writing play here? Do her words refer to a pre-existing relationship between White and her home and work spaces, or does her writing contribute to constructing this relationship?

In this chapter, I piece together a framework for exploring how this quote both underscores and reproduces a violence of gender and 'race' that is often overshadowed by the violence experienced on the physical body. I show that this violence and the materialities it constitutes are enacted through a relationship of space, gender and 'race' that can be explored through the concept of *mobility*, understood as the movement of bodies through social space. Recent academic work (Razack, 2002;

1 A previous version of this chapter was printed as 'Stranger Movements: Working through racialized and gendered mobility in the Canadian context' in the York University Centre for Refugee Studies Working Paper Series No. 4 (2006).

and Mohanram, 1999) has suggested that a spatial understanding of gender and 'race' allows us to better conceptualise their intersection. Importantly, some of this work (Razack, 2002; Mohanram, 1999) argues that the gendered and racialised body is marked by a lessened ability to move and by its belonging to particular spaces.

This chapter brings together these ideas to suggest that if we want to know how mobilities may be gendered and racialised, we may begin by looking at how we write about their 'realities'. I show that by questioning the often-assumed separation between text and materiality, we can better challenge race- and gender-based oppression and its implications for socio-spatial mobility. Thus, the general argument made here is that without examining the constitution of spaces and bodies through meanings of gender and 'race', we may reproduce the oppressions we attempt to destabilise in writing about them.

To develop this argument, I use work by Judith Butler and Frantz Fanon to explore recent work by anti-racist feminist Canadian scholar, Sherene Razack. In the introduction to the edited volume *Race, Space and the Law: Unmapping a White Settler Society* (2002a, 6), Razack argues importantly that in the Canadian context, the destabilisation of meanings of gender and 'race' must include an exploration of their spatial implications. The specific argument made in this chapter is that because Razack assumes an ontological existence for both bodies and space (that is, she suggests that they have a pre-existing and separate being understood as material and not symbolic), her analysis cannot examine, and even further reproduces, the very sexed and racialised constitutions of the spaces and bodies analysed. Finally, this discussion shows that through a focus on the gendered and racialised constitution of the materiality of both bodies and space, we can develop and nuance Razack's argument that the racialised and gendered body is marked by a belonging to particular spaces and a lessened ability to move.

Problematising Razack's analysis

In *Race, Space and the Law*, Sherene Razack (2002a, 5) and other contributors attempt to 'unmap' some of the ways that gender, 'race' and space work together in the English-Canadian context. This unmapping is done in the interest of destabilising the innocence of white European settlers, and of examining continuing ideologies and practices of domination that structure the state (Razack, 2002a, 5). The double purpose of the book is to denaturalise seemingly innocuous geographies by exploring how they are implicated in the production of space, and to destabilise the role of white Canadians and masculinity in this process (Razack, 2002a, 5). Towards these ends, Razack develops a theoretical framework that guides the work of the following chapters. The questions that guide this analysis are complicated and important. She asks: 'What is being imagined or projected onto specific spaces and bodies, and what is being enacted there? Who do white citizens know themselves to be and how much does an identity of dominance rely upon keeping racial Others firmly *in place*? How are people kept in their place? And, finally, how does place become race?' (Razack, 2002a, 5, original emphasis.)

These questions point to the potential and the problems in Razack's analysis. The relationships between space and body, and among gender, 'race' and place, which form the basis for Razack's analysis, are deeply complicated and important, and then she makes the intriguing suggestion that there is a relationship between the 'keeping in place' of some and the 'identity' or domination of others. Despite their importance, these ideas also point to some problematic positions: First, is it accurate to suggest that something is imagined or projected onto a pre-existing material space or body, suggesting that this space or body exists apart from these layered meanings? Is it appropriate to speak of the processes by which white (and non-white) citizens come to 'know themselves'?

Razack's discussion of the body focuses on an argument, following Foucault, that the liberal state is characterised by the separation of the body marked as bourgeois from that body marked as degenerate (2002a, 10–11). This spatial separation, she writes, is secured in the interest of the moral regulation of the body and the state (Razack, 2002a, 11). Razack uses Foucault to argue that throughout the 18th century, the body increasingly became an object and a target of power, and that individuals were 'made' through the micro-processes of discipline (2002a, 11). It is through these processes and technologies of surveillance that the normal social body and the abnormal, spatially segregated, body were produced (Razack, 2002a, 11).

Although the book's focus is on the relationship between racialised and gendered oppression, mobility and space, Razack does not clearly suggest that embodiment, or the production of the body, is itself racialised or gendered. While her understanding of 'race' and gender is not made explicit, her writing suggests that the racialising and gendering of the body happens through other 'identity-making' processes, particularly through the potential for the 'normal' body to move through space (2002a, 13). The bourgeois body of Foucault's analysis is likened to the Cartesian subject in the work of Kathleen Kirby, through the idea that the Enlightenment subject developed alongside the rise of cartographic technology and exploration (Razack, 2002a, 12). This parallel suggests that the figure with the potential to map space is the one who controls it, producing meanings of 'race' and gender by this containment of space (Razack, 2002a, 11). Razack suggests that this mapping subject is also the figure of the imperial man who 'achieves his sense of self through keeping at bay and in place any who would threaten his sense of mastery' (2002a, 12).

For Razack, mobility is a form of identity making. Through the work of Radhika Mohanram and Richard Phillips, Razack argues that the movement from respectable to degenerate space is the process by which the bourgeois subject comes to know himself as such (2002a, 12). She uses Mohanram's work, which explores representations of the black body in various Western texts on identity formation, to show that the black body is always marked and is immobilised through this marking, while in contrast, the white body is characterised by its ability to move freely (Razack, 2002a, 12).

To gender this analysis, Razack looks to Richard Phillips' analysis of 19th century adventure novels for young boys. She focuses on Phillips' argument that a process of subjectivity is enacted through the idea of movement into and back from liminal space, the 'border between civilized and primitive space, [inhabited] by savages whom civilized men vanquish on every turn' (Razack, 2002a, 13). This movement can only be undertaken by the bourgeois white man, and allows this man to identify his spaces and women as civilised, and to know himself to be 'white and in control' (Razack, 2002a, 13). This idea is extended to the Canadian context, where Canadian identity is reaffirmed by middle-class white men's easy physical movements, including those adventures to the inner-city where working-class, racialised and Aboriginal bodies are expected to belong (Razack, 2002b). This framework is later used to show where 'race' and gender were erased from the trial following the murder of Pamela George, an Aboriginal woman who sometimes worked in the sex trade in the Canadian prairie city of Regina (Razack, 2002b, 125). In this haunting analysis, Razack shows how the trial reflected these meanings, focusing on the respectability of the spaces where the accused young middle-class white university men belonged, and the degeneracy and Aboriginality associated with the inner-city where George worked and where these young men reaffirmed their respectable place in the nation (Razack, 2002b, 125).

What remains unclear from Razack's discussion is how the subject and body come into being. As noted, at times Razack refers to the body as produced, at other times it is suggested that a material body exists apart from these 'identity making' processes. Confusion also arises from Razack's discussion of what at times seem to be representations of the body, and at other times seem to be a body existing outside representation. Additionally, throughout the introduction, the relationship between racialisation, gendering processes, and the formation of the subject is unclear, referred to only through the idea of a subject who knows himself through 'identity making' processes that are racialised and gendered.

Despite these tensions, Razack's insights are clearly useful. Is there then a way that we might resolve some of these tensions to determine whether there is indeed a connection between gender, 'race', space and mobility? Although the discussion of the respectable or bourgeois subject is useful, I am unconvinced that it adequately theorises the role of 'race' and gender in the constitution of the subject and body themselves. And if there is a more integrated way that we might theorise about the constitution of the subject and the body through 'race' and gender, then we might be able to better understand how the racialised and gendered body is differently mobile and spatialised from its white masculine counterparts. To better consider the very sexed and gendered production of bodies and space, I begin my analysis by turning to Judith Butler's work on the materiality of the body.

Butler's *Bodies That Matter*

Judith Butler (1993a, 2) integrates an analysis of the production of sex with the production of materiality. She argues that the materialisation, or coming into being, of the body happens through the materialisation of the body's sex (1993a, 2). This materiality of sex is compelled through the regulatory norms of compulsory heterosexuality in a performative fashion, that is, by producing that which is being named (1993a, 2; 1993b, 13). Although the body is posited as existing before language, she argues, 'this signifying act delimits and contours the body that it then claims to find prior to any and all signification' (1993b, 30). In other words, the body's materiality is woven into signification and so does not happen outside language (1993b, 31). For Butler, materialisation is never complete, but over time is stabilised in and through the appearance of what we understand to be matter: 'the effect of boundary, fixity, and surface' (1993a, 9).

The idea of performativity, taken from speech act theory, depends on a notion of power that is not the act of an originating will (Butler, 1993a, 13). Through a revision of the grammar of action, power is understood as that which is produced in and through its effects, which are themselves power (Butler, 1993b, 35). Thus, we are not dealing with a subject who acts or power that acts on a subject. Butler writes: '"Materiality" designates a certain effect of power, or rather, *is*, power in its formative or constituting effects. Insofar as power operates successfully by constituting an object domain, a field of intelligibility, as a taken-for-granted ontology, its material effects are taken as material data or primary givens' (Butler, 1993b, 34–5). The moment in which matter appears as if it exists outside the workings of power or discourse, as an irreducible ground of meaning, is the moment when this discourse is most effective (Butler, 1993b, 35).

These ideas can help us to reformulate Razack's hypothesis that the body that is racialised and gendered is not freely mobile and belongs to particular spaces. First, Butler's understanding suggests that by considering processes that make 'identities' of 'race' and gender *on* the body, we further reproduce the gendered constitution of the body itself. Rather than understanding how gendered and racialised identities are made separately from the body, the making of these 'identities', which happens through the constitution of sex itself in relation to compulsory heterosexuality, is the very making of the body itself. Further, we can no longer speak of anything outside discourse, and when we speak of the body we are always speaking about something produced in its representation. Butler's ideas complicate the relationship of separation that seems to be suggested in Razack's discussion of matter and its representation, and show how all materialisation happens through the constitution of sexual meaning.

But what does all this mean for the idea of the 'subject'? Butler argues that subjects of action come to appear as an effect of this materialisation through sex (Butler, 1992, 9). Butler again takes from Foucault, arguing that the subject is the effect of a chain of actions, a genealogy that is erased when the subject is understood as the foundation of action (Butler, 1992, 12). The subject is not the point of origin of action, but is itself the effect of various acts, which are unpredictable and indirect (Butler, 1992, 12). This constitution of the subject through the materialisation of sex is then covered over through the supposed autonomy associated with the subject (Butler, 1992, 12). Although the subject is posited as an actor within an external sphere of social relations, these relations are the very process, then concealed, of the subject's constitution (Butler, 1992, 12). This constitution of the subject requires, for Butler, an identification with norms of sex through a repudiation, forming an abjected outside which must then be covered over and disavowed (Butler, 1993a, 3). If not, this would threaten to reveal the 'self-grounding presumptions' that are at the basis of the sexed subject (Butler, 1993a, 3).

Butler's approach raises questions about Razack's focus on the subject's self-awareness. If the subject is produced as an effect of the materialisation of the body, and if the actions of the subject cover over its constitution, then we may not get to the roots of oppressive 'identity' by focusing on the process by which this subject 'knows himself'. Instead, if we assign the subject this agency that then covers over the constitution of that subject through sex (and 'race', as we see below), then those norms that compel such a constitution are further reproduced.

It is important to note that at the start of *Bodies That Matter*, Butler quotes Gayatri Chakravorty Spivak's argument that the most important critique made by deconstruction 'is the critique of something that is extremely useful, something without which we cannot do anything' (Spivak in Butler, 1993, 28). Although her work problematises the meanings of the 'subject' or the 'body', she does not suggest that the concepts should not be studied or used (for example, see Butler 1993, 29, for a discussion of the concepts of 'woman' and 'materiality'). Instead, it is important to examine how these ideas materialise, and what is grounded through this materialisation. Razack's analysis of mobility suggests that we focus on how racialised and gendered bodies are made intelligible, and in which ways their intelligibility is bound up with particular spaces.

The possibilities for resistance are located within this process of constitution. Butler's understanding of resistance requires that the subject and its agency are not considered external to the field of power, nor that the subject is the effect of one act of power (Butler, 1992, 13). Rather, when the subject is understood as continually being subjected and constituted, when its materiality is understood as the effect of reiterations that congeal over time to produce an appearance of stability, the possibilities for resignification are left open (Butler, 1992, 13). Importantly, Butler argues that this materialisation of the sexed body is the effect of a set of actions that are mobilised through, but not compliant with, the regulatory law; thus, the performative constitution of the body is always engaged in a reshaping of meaning (1993a, 12).

One of Butler's critical revisions in *Bodies That Matter* (1993a) is the role played by racialisation in the regulation of sex and gender. In this work, Butler revisits her earlier discussions of performativity and the construction of sex and gender to suggest that it is not only the regulatory power of compulsory heterosexuality, but also taboos against miscegenation that structure the materialisation of bodies. Although Butler brings attention to this oversight and argues that the heterosexual matrix that compels the materialisation of sex is always racialised, she does not adequately theorise 'race' in her work. Despite this omission, Butler's analysis points to how a truly intersectional analysis might be done. In the following section, I turn to work by Frantz Fanon to piece together such a framework.

Fanon and racialised mobility

Racialization for Fanon, as for Razack, does not happen by virtue of having a black body, but can be understood as taking place through what Razack might call 'identity-making processes'. What is different in Fanon's analysis is that 'identity' is not merely assumed or played out on the surface of the body, but, as is true for sex in Butler's account, racialisation itself constitutes the body (Fanon, 1967, 112). Fanon shows how the black body is brought into being in his often-cited 'Look, a Negro!' sequence, described briefly below. What is important to note for my analysis is how Fanon understands this embodiment to be played out in terms of the subject, the body, 'race', and space, and how this analysis relates to my earlier discussion of Butler and the sexed materialisation of the body. Social space enters into Fanon's analysis in interesting ways; first, as that which provides the context for embodiment, and secondly, in the relationship of the racialised body and social space. I use Fanon's work to show how we might better understand how bodies and spaces produce one another, and how the racialised body is constituted through immobility.

In Chapter Five of *Black Skin, White Masks* (1967), Fanon describes the movement of pre-1940 Antilleans from Martinique to colonial France. The Antillean 'at home' is immersed in French discourse, including French school curriculum, French magazines and stories, and French history, and so comes to identify with whiteness as the norm (Fanon, 1967, 146). As whiteness is associated with purity and morality, and blackness is associated with immorality and sin, the Antillean, to see himself as moral, must maintain a psychic image of himself that is white (Fanon, 1967, 192). Fanon argues that in this way the collective unconscious of pre-1940 Antilles is white, and the black man is understood as the African, particularly the Senegalese (1967, 148). This understanding is troubled through maturity, and although the Antillean may identify himself as Negro, he still must maintain a moral identification with whiteness (Fanon, 1967, 192–3).

Once he is in France, the Antillean is confronted with the gaze of the white man ('Look, a Negro!'), and is confronted with his blackness (Fanon, 1967, 112). In becoming black, the Antillean realizes that he, and not just the African, is represented by the 'myth of blackness', constructed through white images and histories of the Negro (Fanon, 1967, 112, 150). This marks the zebra striping of the mind – the white unconscious is too stable for an easy displacement and the Antillean is left fragmented and 'forever in combat with his own image' (Fanon, 1967, 194). As Ann Pellegrini suggests, this white gaze is most successful as it becomes internalised (1997, 92). This gaze also results in a trauma to the bodily schema, or the 'lived body by and through which one takes up the world' (Merleau-Ponty in Sullivan, 2004, 13). When the black Antillean is confronted with the white world, this bodily schema falls away to show what lies underneath, the existence of what Fanon calls the historico-racial schema formed through representations of blackness in discourse (Fanon, 1967, 111).

Although blackness comes into being in relation to whiteness, the converse is not also true and any possibility of black subjectivity is precluded (Fuss, 1994, 22). As Diana Fuss suggests, within white colonial discourse, the Other, always blackness, is used to maintain colonial selfhood (Fuss, 1994, 22). Although whiteness may be 'the other' to blackness, it can never be the Other that is essential to subjectivity (Fuss, 1994, 22). Rather, blackness is produced as a sort of illusory Other through a subject/other relationship that is actually an economy of the same, as illustrated by Fanon's discussion of the myth of blackness. The blackness that becomes the Antillean is made up of white myths: 'I discovered my blackness, my ethnic characteristics; and I was battered down by tom-toms, cannibalism, intellectual deficiency, fetishism, racial defects, slave ships, and above all else, above all: "Sho' good eatin"' (1967,

112). Fanon's work performs a further deconstruction of colonial subjectivity by suggesting that if the Antillean forms the exteriority of the colonial interiority, then the very humanness of the coloniser, defined against the non-humanness of blackness (which is only produced, falsely, in relation to whiteness!) comes into question (Fuss, 1994, 23). Fuss describes this as the double command of the colonizer: 'Be like me, don't be like me' (Fuss, 1994, 23).

Fanon's account of the preclusion of black subjectivity problematises and supplements Razack's framework on a number of counts. First, this analysis suggests that there are a number of very important links between 'race', embodiment, subjectivity, and mobility to be considered alongside our earlier discussion of gender. However, it also suggests that when Razack (2002a, 5) asks: 'Who do white citizens know themselves to be and how much does an identity of dominance rely upon keeping racial Others firmly *in place*?' she poses a very complicated question, one that, as discussed above in relation to Butler, reproduces the illusory effect of a thinking and acting subject. Further, Fanon suggests that the racialised body becomes neither other nor subject, but simply an object. Whiteness, the 'transcendental signifier', is not understood as 'not-black', but is self-reproducing and works as its own Other (Fuss, 1994, 144). This suggests the importance of moving away from questions of how the subject knows himself and asking instead how this subject comes to appear through the 'keeping in place' of racialised bodies, and how this keeping in place is achieved.

Racialisation in Fanon's account is the embodiment of blackness, which is embodiment itself for the black man, unrecognisable without his blackness in the French context. This is, at the same time, the process by which the body comes into being and the process by which the black man's subjectivity is denied. The black man cannot attain a disembodied subjectivity but remains characterised, only a 'black man' and never a 'man', in a heightened state of embodiment. Referring back to this becoming of blackness, Pellegrini writes: 'This inscription surfaces on and as the racialized body; the oppressed loses all claim to individual "being" and becomes instead a representative type' (1997, 92).

Not only does embodiment happen in relation to social space, but in this case, racialisation happens in relation to bodies that produce the social space of colonial France, and is also the process by which the racialised body is marked as one that cannot move freely in space because of the trauma to its bodily-schema (Fanon, 1967, 112). As Fuss acknowledges:

> Space operates as one of the chief signifiers of racial difference here: under colonial rule, freedom of movement (psychical and social) becomes a white prerogative. Forced to occupy, in a white racial phantasm, the static ontological space of the timeless 'primitive', the black man is disenfranchised of his very subjectivity. Denied entry into the alterity that underwrites subjectivity, the black man, Fanon implies, is sealed instead into a 'crushing objecthood'. (Fuss, 1994, 21)

This hyper-embodiment can also be understood as a sort of alienation of the black man from the black body (Dalal, 2002, 98). Fanon argues that the now-racialized person is left in a state of seeking approval and definition from the white world, which cannot come from the non-existent Other (Sullivan, 2004, 15; Fanon, 1967,

154). The black body only comes to take on meaning in a social space where the white body is presumed to be *the* body.

If Fanon's work is applicable to contemporary contexts,[2] then a number of things might be suggested. First, in light of Butler's understanding of embodiment, an 'identity' of whiteness might itself never be fully achieved, but might be better understood as needing continual reaffirmation through encounters with an Other produced in its own image, as is highlighted in Razack's thoughts on the movement of white, middle-class men's bodies into 'degenerate' or racialised space. Secondly, although Butler's arguments effectively show that subjects are merely an effect of materialisation, much academic and non-academic writing, including anti-racist feminist writing, continues to conceptualise bodies as belonging to acting subjects.

Fanon's work, taken together with Razack's suggestion that people of colour and Aboriginal people in Canada continue to be excluded from subjectivity, suggests that those who are excluded are not merely an Other, but are, to some extent, hyper-embodied as a representative racialised type. Racialised bodies become intelligible in part through a process that forms part of white subjectivity, itself an illusory process. It is for this reason that the racialised body falls into a sort of representative position that is the basis for its intelligibility. The hyper-embodiment of the body of colour in this context does not translate into immobility, but characterises the body primarily as a material body. An effect of this marking is that the mobility of the body is not the free and transformative mobility of the Subject, but always continues to ground racialised and gendered meanings, as may be clear in a consideration of the 'nomad' or the black athlete or dancer.[3]

Again, these arguments leave room for resistance. Fanon cautions that, despite the representation of blackness in white discourse, something remains elusive about the black man, some mystery or secret that the black man keeps back from his white counterpart (1967, 128). Radhika Mohanram (1999, 27) relates this mystery to the nature of racialised representations, particularly with respect to what Fanon understands as the necessity to deny the existence of 'Negroes' in favour of representations of 'the Negro' (Fanon, 1967, 127, as one example). Mohanram writes, 'the supplementary nature of Fanon's blackness must be excluded for the internal coherence of a system divided into blacks and whites to function, a system defined from the vantage point of whiteness' (1999, 27). This suggests that the racialised body is not without possibilities for resistance.

2 It is important to consider to what extent Fanon's work might be generalised to other contexts. It is clear that Fanon's analysis speaks primarily about black men. However, if we consider that feminist works such as Luce Irigaray's *This Sex Which Is Not One* (1985) follow similar lines to suggest that women are also inside/outside the process of subjectivity constituted through Self-Self relations, then it is possible to argue that Fanon's work holds relevance for those 'identities' of Other, that include, Razack (2002a, 6) argues, representations of people of colour and Aboriginal peoples in a Canadian context.

3 Tim Cresswell's insightful comments allowed me to further develop these ideas.

Moving forward

These ideas about racialised and gendered mobility suggest that it is important to consider how meanings of 'race' and gender are reproduced in writing, but also in writing that seeks to expose these meanings. In this chapter, a critique of Razack's work was performed to better show how this analysis of writing might happen and how this analysis may be applied to contemporary contexts. The work of Butler, Fanon, and other post-colonial scholars was used to argue that any discussion of racialized and gendered mobility must very carefully consider how subjects, bodies and spaces are constituted through meanings of 'race' and gender.

This analysis allows us to reconsider the case of Liana White and other missing and murdered women of Edmonton.[4] When read in relation to this analysis, it is possible to suggest that Paula Simons' writing both refers to and reproduces meanings and 'realities' of who belongs where within Canadian national discourse and spaces. Throughout the media coverage of the Edmonton sex trade murders, particular areas of inner city space were identified as the zones from where mostly Aboriginal women worked and disappeared. In Simons' words (2005, B1), White's properly vulnerable femininity and respectability due to pregnancy, motherhood, and neo-natal work is closely tied to her belonging to the suburbs, areas that Razack shows to be associated with whiteness and normalcy (2002b, 140).

To better analyse this violence, we may look at how the relationships between text and materiality are at play. This removal of women is made sense of in how we write about it. In referring to these spaces, the place of white women and Aboriginal women is reaffirmed; proper women citizens come into being through their belonging to spaces of normal femininity and respectability and are out of place in inner-city spaces where some other women belong. Simons' words both refer to and further constitute these bodies and spaces in violent ways that keep some in place and mark others as subjects.

This discussion suggests the importance of not only revising our understandings of the body and subject when writing about racialised or gendered mobility, but also conceptualisations of space. The questions put to Razack's understanding of the body and the subject, through Butler's work on the materiality of the body, equally pose questions for her understanding of space. What does this understanding of the body suggest for an understanding of space? How is social space produced, and is this production also caught up in the production of race and gender? How can we challenge these relationships to make space for all women to move freely in public spaces, for example? These questions underscore the importance of future work, for if social space may be conceptualised as continually reproduced through and as an effect of gendered and racialised embodiment, an effect that then further constitutes the bodies that produce it in a circular relationship, then its constitution alongside embodiment holds further potential for resistance and disruption.

---

4 These incidences reflect a trend of violence against women, repeated not just in the cities of Edmonton and Regina, but also in Vancouver, Canada, and recently, Ipswich, UK. In all of these cases, women working in the sex trade are removed from inner-city public space, murdered, and abandoned in rural areas outside the city.

References

Butler, J. (1992), 'Contingent Foundations: Feminism and the Question of 'Postmodernism.' In Judith Butler and Joan W. Scott (eds) *Feminists Theorize the Political* (New York and London: Routledge).

Butler, J. (1993a), 'Introduction.' In *Bodies That Matter: On the Discursive Limits of 'Sex'* (New York and London: Routledge).

Butler, J. (1993b), 'Bodies That Matter.' In *Bodies That Matter: On the Discursive Limits of 'Sex'* (New York and London: Routledge).

Dalal, F. (2002), *Race, Colour and the Processes of Racialization: New Perspectives from Group Analysis, Psychoanalysis, and Sociology* (Hove and New York: Brunner-Routledge).

Fanon, F. (1967), *Black Skin, White Masks*. Translated by Charles Lam Markmann (New York: Grove Press).

Fuss, D. (1994), 'Interior Colonies + Colonial History and Racism - Fanon, Frantz and the Politics of Identification.' *Diacritics-a Review of Contemporary Criticism* 24 (2/3), 20–42.

Grosz, E. (1995), 'Space, Time and Bodies.' In *Space, Time and Perversion: Essays on the Politics of Bodies* (New York and London: Routledge).

Harding, K. 'The 'Killing Fields' of Edmonton.' *Globe and Mail*, A1, 2 August 2005. Retrieved through Factiva news database, document number: GLOB000020050802el820002n. Accessed 29 August 2005.

Irigaray, L. (1985), *This Sex Which Is Not One*. Translated by Catherine Porter (Ithaca: Cornell University Press).

Mohanram, R. (1999), *Black Body: Women, Colonialism and Space* (Minneapolis: University of Minnesota Press).

Pellegrini, A. (1997), *Performance Anxieties: Staging Psychoanalysis, Staging Race* (New York: Routledge).

Razack, S. (ed.) (2002), *Race, Space and the Law: Unmapping a White Settler Society* (Toronto: Between the Lines).

Razack, S. (2002a), 'Introduction: When Place Becomes Race,' In Sherene Razack (ed.) *Race, Space and the Law: Unmapping a White Settler Society* (Toronto: Between the Lines).

Razack, S. (2002b), 'Gendered Racial Violence and Spatialized Justice: The Murder of Pamela George.' In Sherene Razack (ed.) *Race, Space and the Law: Unmapping a White Settler Society* (Toronto: Between the Lines).

Simons, P. (2005), 'The 'Show' is Over, Now it's Time to Mourn: Liana Deserved a Better Ending to her Life's Story.' *The Edmonton Journal*, B1, 21 July 2005. Retrieved through Factiva news database, document number EDJR000020050721e17l00001m. Accessed 29 August 2005.

Sullivan, S. (2004), 'Ethical Slippages, Shattered Horizons, and the Zebra Striping of the Unconscious: Fanon on Social, Bodily and Psychical Space.' *Philosophy and Geography* 7 (1), 9–24. Retrieved 27 September 2004, from Ingenta database. Available through Carleton University Libraries Website.

Chapter 4

Motherhood, Risk and Everyday Mobilities

Lesley Murray

> Karen begins her day on the 6:50 a.m. bus that fortuitously stops just outside her front door. Her workplace is just 30 minutes away by bus, and she does not have to be at work until 9:30 a.m., but every weekday she escorts her son to secondary school, taking two buses to get there and another one back to her place of work. Karen began accompanying her son to school following his experiences of being bullied on the bus at the beginning of his first term in secondary education. Meanwhile, Cheryl, whose son attends the same school, leaves home at about 7:30 a.m. and drives both her children to their respective schools, about twenty minutes apart and in the opposite direction from her workplace. She arrives just before her shift begins and leaves when it ends at 1:30 p.m. She usually drives straight to her son's school to ensure that she gets a parking space nearby. She returns home at 3:45 p.m., having spent nearly half of her day in the car escorting her children to and from school. (Murray, forthcoming.[1])

There are indications that motherhood is becoming increasingly constructed by the hypermobilities[2] that are a part of Karen and Cheryl's everyday lives. The multidimensionality of motherhood and the complex landscapes of risk (Valentine, 2004) that both mothers and children negotiate every day (Tulloch and Lupton, 2003) are precursors to this relatively contemporary characterisation of mobility. The resulting patterns of mobility are both gendered and generational (Murray, forthcoming), with this chapter, which is concerned primarily with motherhood, centred on the former. Although gender has been recognised as a significant determinant of transport exclusion in the UK since the 1970s (Hillman, 1976), with a range of research, both academic and policy orientated, focusing on women and transport (Pickup, 1988; Tivers, 1988; Hamilton and Jenkins, 2000; Greater London Council, 1985; London Research Centre, 1998; Law, 1999; Domosh and Seager, 2001), there are few studies that look specifically at motherhood and mobility from

1 This illustration of the everyday mobilities of two mothers is from a study of the risk landscapes and mobilities of the journey to school undertaken in Brighton, UK between 2005 and 2006. Twenty mothers and their children were interviewed and their journeys recorded by video camera.

2 Hypermobility is used by Adams (1999; 2005) to describe the grave consequences for society of unsustainable and deleterious increases in mobility. The term is developed in this chapter to characterise the negative impacts of increasing mobility for mothers due to their gender role.

a theoretical perspective. Recent contributions from the new 'mobilities paradigm' (Sheller and Urry, 2006) focus on the gendering of the car (Sheller, 2004) and the impact of car culture on gendered spaces (Sheller and Urry, 2003).

This chapter explores the concepts of motherhood and risk through their interrelations, and argues that they not only construct everyday mobilities, but that they are co-constructed within an increasingly mobile society. Mothering culture and risk experience are dependent on mobilities, as the mobile society is shaped by gender and risk. The chapter also argues that cultures of mothering are not only determined by ideology but through everyday risk and mobility practices.

The term 'mother' is both problematic and dynamic in all its representations. In its broadest sense, motherhood is the 'context in which mothering takes place and is experienced' (Miller, 2005, 3). It should be stressed that the notion of 'mother' is itself fluid and contested. Mothers are a diverse group, with factors such as class, race, sexuality, disability and ethnicity contributing to both gendered identities and practices of mothering (Gillies, 2006; McDowell, 1999; Miller, 2005). For Miller, being a mother is always more than 'playing a part' (Goffman, 1969, 28 cited in Miller, 2005, 14), even though it is socially and culturally embedded, and the biological act of giving birth and the central responsibility of motherhood to meet the bodily needs of dependents is central to its construction. A full understanding of motherhood requires consideration of its spatiality (Holloway and Valentine, 2000) and in particular its mobility. In addition, both the construction of motherhood ideology and mothering practices, including mobile practices, are determined by risk.

Studies of risk and gender have shown that motherhood, as well as gender, class, age and ethnicity, can have a particular influence on risk perception and experience (Douglas, 1986; Lupton, 1999; Mythen, 2004; Slovic, 2000). As Lupton (1999) argues, risk taking is associated with the performance of masculinity that conforms to cultural norms, while in contrast, risk aversion is associated with dominant femininity and the everyday responsibilities of motherhood, although there are significant examples of contrasting gendered behaviour. Lupton argues that these behavioural characteristics are acculturated from an early age. Douglas also differentiates between the risk perceptions of men and women, particularly mothers, arguing that women overemphasise their vulnerability to crime as they are 'socialised into high risk awareness' (Douglas, 1986, 70).

Like risk and mobility, motherhood is both a socially constructed and embodied position, which can be conceptualised through a range of theoretical levels. The co-construction of motherhood, risk and mobility can be examined at a number of theoretical levels, from local to global. This chapter begins by examining the concepts on a macro theoretical scale, based on global notions of 'normality' and 'good' parenting.

Dominant ideology, individualisation and experts

Despite ideological notions of a 'good' 'nuclear' family (Valentine, 2004), the structure of the family in western countries is changing (Beck and Beck-Gernsheim,

1995; Silva and Smart, 1999; Valentine, 1997; 2004, Clark, 1996), with an increase in single parenthood (Duncan and Edwards, 1999) and a corresponding feminisation of childhood, because single parents are predominantly mothers rather than fathers (Silva and Smart, 1999; Valentine, 2004; Wyness, 2000). As a result, a significant proportion of both mothers and children are 'out of place' in the context of these changes. Phoenix and Woolett (1991) argue that the idealised construction of the 'good mother' is dependent on 'dominant ideologies', which means that working class mothers, young mothers, single parents, mothers from minority ethnic groups and disabled mothers are considered to be deviant from norms of proper child rearing (Phoenix and Woolett, 1991; Chase and Rogers, 2001; Valentine, 2004; Connolly, 2000). Phoenix and Woolett (1991) contextualise motherhood within the social structure, and argue that the state protects these dominant ideologies through processes of surveillance and intervention.

In broad terms, two conflicting 'dominant ideologies' of motherhood can be identified based on encouraging women to return to work after they have children through policies such as New Deal[3] on the one hand, and an adherence to traditional moral values that prescribe a full-time home-keeping, child-rearing role for mothers on the other (Miller, 2005; Skinner, 2005). Motherhood at an everyday level thus becomes associated with variations from established global norms of good parenting. In addition, since the early twentieth century and particularly the 1950s and 1960s, with their emergent 'expert' discourse, there has been considerable pressure on parents to nurture their children psychologically (Beck and Beck-Gernsheim, 1995, Kelley et al., 1997; Lewis, 1980; Marshall, 1991; Urwin, 1985). The nature and targeting of expert guidance, particularly through media amplification (Beck and Beck-Gernsheim, 1995; Weaver, Carter and Stanko, 2000) and the degree to which parents feel pressured to adopt dominant cultures of parenting is dependent on a range of cultural characteristics and is particularly influenced by class.

It is argued here that dominant ideologies and the legitimisation of expert knowledge is based on discourses of risk, and that risk plays a profound role in establishing cultures of mothering, particularly mobile mothering. Social scientists informed by the work of Michel Foucault (Fotel and Thomsen, 2004), argue that state control through allocation of risk is based on expert knowledge, particularly the process of normalisation. In this way parents are coerced into control and surveillance of their children, resulting in their loss of independent mobility and an increase in parental escort trips (see Hillman et al., 1990; Joshi and MacLean, 1995; Fotel and Thomsen, 2004). The state therefore denies individual agency through the imposition of expert-driven management of risk.

In contrast, Beck and Beck-Gernsheim (1995) argue that within the context of expert notions of parenting, motherhood is increasingly defined by the late modern process of individualisation, a process that recognises an individual's agency in determining her own mothering and mobility outcomes. The identities of mothers become bound up with notions of increasing emotional investment in children (Beck and Beck-Gernsheim, 1995; Jenks, 1996; Valentine, 1997). Ideologies that seek to

3 *New Deal* is a UK government programme aimed at encouraging unemployed people back to work. There are specific incentives for single parents.

protect established notions of childhood as an investment in the future promote this individualised emotional investment (Hendrick, 1994; 2003; Wyness, 2000). As discussed later, this investment in children can have a direct impact on mobility options and practices. However, the extent to which this is a universal characteristic of motherhood is contested.

Chase and Rogers (2001) argue that identity enhancement, the enrichment of personal identity through reflexive processes in motherhood, is a characteristic of middle-class mothering. For working-class mothers, it is argued, the practical issues of reduced choice and the constraints of motherhood are more important. In line with this, Gillies (2006, 284) argues that the notion of 'emotional capital',[4] with its emphasis on positive outcomes, is less meaningful to working-class parents. She asserts that in working-class households, mothers make significant investments on an emotional level, but that this is more likely to be based around risk minimisation. For middle-class parents, therefore, it is argued that risks are less tangible and more likely associated with reflexive practices. Miller (2005) argues that reflexivity needs to be understood in a structural context, linked to culture, class or race, and that external structures still restrict mothering practices. In addition, self-reflexivity is not universally experienced in relation to material and structural contexts, as the middle classes have the resources to self-reflect and establish identities in this way.

The processes of reflexive modernisation and individualisation,[5] which are a part of Beck's (1992) 'risk society', have been criticised by motherhood (Lewis, 2005; Miller, 2005) and risk theorists (Adams, 1995; Lupton, 1999; Mythen, 2004; Scott, 2000; Wynne, 1998). Beck argues that reflexivity liberates women from their gender-ascribed roles as 'men and women are released from traditional forms and ascribed roles in the search for a 'life of their own'' (Beck, 1992, 105). Beck's theory is more applicable on a macro level, to technological risks that are less relevant to people's everyday lives (Adams, 1995; Lupton, 1999; Scott, 2000; Tulloch and Lupton, 2003). It is through socio-cultural risk theory that everyday risk can be situated within dominant ideologies. Theory related to the allocation of blame in society is particularly useful in explaining cultures of mothering.

Allocation of blame

The allocation of 'blame' as a result of expert discourse of risk was put forward by Mary Douglas (Douglas and Wildavsky, 1983; Douglas, 1986). Douglas' cultural theory of risk asserts that risk perception is constructed on a number of levels, including an individual level, where personal solutions are applied to individually perceived risks, and a collective level, by scapegoating vulnerable groups. Douglas (1986) argues that a community is more likely to exclude someone who takes risks.

---

4 Drawing from previous discussions of this term, Gillies (2006) broadens its definition to describe the pool of emotional resources available for investment in children's future wellbeing.

5 Beck's individualisation is based on a new focus on the self as the primary unit of identity formation, which results from a breakdown of various groupings of individuals by community, class, and gender.

This could include, for example, mothers who are out of step with local practices around children's independent mobility. Predictions of whether cultures are risk-averse or risk-inclined can be based on what or who is at risk, and the levels of solidarity and structure maintained within the community (Douglas, 1986). Douglas' notion of 'dangerousness', a term describing those who seek out risk, is attached to marginalised groups, including mothers, and particularly those whose cultures and practices fall outside established norms. This marginalisation legitimises spatial as well as social control.

Douglas (1986) cited examples of expert discourse around risk, such as attachment theory developed in the 1950s, which she argues had a direct impact on mothers' mobility by reducing mothers' ability to accept employment. Although an example of blame directed at middle-class career-orientated mothers, blame is most often directed towards marginalised mothers and those who are considered to fall short of constructions of 'good' parenting, and is dependent on a range of social characteristics including class, race and sexual orientation (Phoenix and Woolett, 1991; Chase and Rogers, 2001; Valentine, 2004; Connolly, 2000).[6] This is particularly the case in a political climate that seeks out scapegoats for social deviance with links between parenting practices and anti-social behaviour (Gillies, 2006). The attribution of blame is therefore recognisable on both a global and local level, through expert discourses and local cultures of 'good' mothering. It is argued here, however, that in the context of everyday mobility, it is not only the impact of global ideologies but their interpretation within everyday social experience that is formative.

Social contexts of mothering

Duncan and Edwards (1999) propose that the position of mother is based on both individual identity and social context, which reflects dominant cultures and beliefs. They argue that mothers are coerced into their social position and constrained by norms of good parenting, as they strive towards the 'moral imperative' of meeting their children's needs. Mothering ideals, they argue, are imposed by legislation, professional practice or neighbourhood networks, but can vary according to the particular social context of motherhood. However, mothers themselves adopt particular 'gendered moral rationalities' based on being primarily a mother, as both a mother and worker combined, or as primarily a worker. Different gendered moral rationalities will determine different cultures of mothering and mobility practices. However, as Miller argues, the notion of the 'good' ambivalent mother combining work outside the home with childcare refers mainly to a particular group of white middle-class women and is less relevant to other groups (Miller, 2005). This has particular implications for mobility. As mothering practices vary by social status, so too will mobility and mobility choices.

Miller (2005) argues that it is the social contexts of mothering that determine mothering culture, as opposed to the singular notion of dominant western ideologies,

6 See for example Connolly's (2000) study of homeless mothers in the USA.

which ignore the diversity of women's experiences. Normative mothering is dependent on cultural beliefs and meanings and how these are interpreted by other cultures. She maintains that women exercise agency within the context of dominant ideologies. For Miller (2005, 3), mothering is about 'personal, individual experiences that women have in meeting the needs of and being responsible for their dependent children'. In turn these individualised experiences take place in the context of ideologies that determine the characteristics and behaviour of a 'good' mother, whether or not they are consistent with the realities of motherhood (Connolly, 2000). This notion of the 'good' mother is bound up in mobility and risk, the latter through the allocation of blame. The 'good' mother has access to and adopts particular mobility practices, which in turn contribute to cultures of good mothering.

Discourses of everyday risk and mothering

Everyday meanings and identities of motherhood are bound up with notions of good parenting, which are themselves inextricably linked to risk experience and mobility. This is particularly the case with children deemed to be in risky situations, with blame not only from the local community but also from mothers who blame themselves, and with mobility restrictions imposed as a result (Valentine, 1997, 2004). Everyday discourse and the development of local cultures of mothering can be empowering, as will be discussed later in the chapter, but they can also represent an expanding network of blame (Dyck, 1990; Valentine, 1997, 2004) necessitating greater sensitivity to risk-taking. For middle-class parents in particular, the desire to network may be bound up in issues of emotional capital with pressures to maximise their children's social potential (Gillies, 2006). In doing so, mothers increase social interaction and become increasingly mobile (Barker, 2003).

Backett-Milburn and Harden (2004) emphasise the importance of the everyday 'mundane' negotiation of risk in determining parenting practices, including mobility practices. They argue that risk is minimised through 'the manipulation of space, place and people' (Backett-Milburn and Harden, 2004, 435). Everyday risk is thus experienced through mobility. They argue that this is a dynamic process where 'experts' are created and re-created in time and space, and where risk is negotiated and renegotiated. As argued, the risk minimisation process, which is also mobility minimising, emerges from cultures of blame, and can paradoxically lead to accusations of paranoia. Everyday discourses around risk and children contribute to the construction of a mothering culture based on expert-induced uncertainty and blame, in turn leading to increased parental fears for their children in space. As discussed, this can lead to a significant impact on both mothers' and children's mobility, as there is an increased dependence on escort trips. The escalation of fear then leads to the creation of notions of parental paranoia (Hillman, 1993; Furedi, 2002), which are considered to be symptomatic of our 'culture of fear' (Furedi, 2002).

However, the paranoid parent discourse can be critiqued for its lack of attention to the complex interaction of motherhood, risk and mobility on an everyday level (Jackson and Scott, 1999; Pain, 2003). This complex interaction includes the consideration of not only mothers' perceptions of risk, but also their everyday risk experience, which is bound up with their personal biography and spatiality (Tulloch, 1999; Tulloch and Lupton, 2003). Tulloch (1999) argues that the risk-aversion of women is based not on irrational perceptions but on their experiences of risk and their expertise in processing this experience. He undertook a major 'fear of crime' research project, which showed that women adopt coping strategies to maximise their mobility in public spaces and that these strategies were based on their experiences. Tulloch argues that charges of irrationality directed at women are based on 'decontextualised' and 'expert-driven' research, which poses questions that are 'too ambiguous and hypothetical' (Tulloch, 1999, 186). Risk experience can therefore be seen as fluid and dependent on everyday mobilities in socio-cultural contexts but also in time, place and space.

Risk and motherhood in mobile space

Much of the theory relating to risk and motherhood, although recognising the importance of social, cultural and political processes, fails to contextualise these processes in space. It therefore underestimates the impact of mobility, and particularly embodied mobility, on these concepts. As McDowell (1999, 5) argues '…space and place are gendered and sexed, and gender relations are 'spaced'', illustrating this with examples of the 'embodied knowledge' of young mothers in working-class communities in Britain and the USA. For McDowell (1999), the term embodiment encompasses the fluidity and flexibility of the socially constructed nature of the body, which for women can be the root of their subordination. Holloway and Valentine (2000) illustrate how locally negotiated ideas of good mothering and negotiation of boundaries are spatially defined. They argue that both homes and schools are porous rather than bounded, and that they are shaped by local and global processes, including children's social networks. Mothering cultures themselves are spatial and mobile, as spatiality defines cultures of mothering through discourse (Dowling, 1999).

Similarly, Dyck (1996, 126), in her study of women in a residential suburb of Vancouver, found that women's identities as mothers are 'embedded in the notion of safe space, which is defined and negotiated by the women in their daily mothering work'. Dyck (1990) related mothering practices to women's social interactions and the political and economic context in which decisions about their children's use of space are made. Women negotiate 'good' mothering through these everyday social interactions, adopting strategies to avoid blame. This occurs through everyday discourse with friends and acquaintances and within different levels of 'expert' intervention, from regular meetings with child health and educational professionals to central government policy and information.

This mothering identity based on established norms of good parenting can be associated with particular spatial contexts and particularly that of the private car. In her study of suburban mothers, Dowling (1999) found that the car was instrumental in constructing mothering identities based on parental aspirations for their children. She found that the use of cars amongst middle-class parents in suburbia enabled parents to maximize parenting abilities. They used the car to maintain the prescribed image of 'good parents'. The person-car hybrid (Urry, 2000) in this instance becomes the culturally defined 'good' mother. Urry argues that this hybrid is based not only on changing patterns of individual behaviour but is reflective of social life, a society constructed through different mobilities. He argues that automobilisation constructs social life through coercion as it imposes 'intense flexibility' (Urry, 2000, 191). The automobile mother is therefore forced to adopt multiple roles as flexibility is imposed by a car-driven society. This is less applicable, however, to the localisation of mothering discussed previously and it is therefore difficult to explore the small-scale everyday mobilities that characterise much of mothers' mobility without referring to relative mobility in public and private space.

The mobility of the public and the private

The public/private dichotomy is important here as it represents the socio-cultural gendering of urban space and the impact of risk and mobility on this process. It demonstrates how the concept of motherhood and mobility are co-constructed and the different theoretical approaches to gender and mobility. The divide is based on a gendering of space within a patriarchal society, with women controlled through the domesticated private space of the home (Massey, 1994; McDowell, 1999). Urbanisation in the early modernising period gave rise to the potential for increased mobility, particularly for middle-class women, including the female flaneur (Law, 1999), blurring the boundary between private and public space and threatening patriarchal control (Massey, 1994; McDowell, 1999; Domosh and Seager, 2001). The private-public divide is definitively linked to mobility, whether through the localisation of women's everyday lives (Little et al., 1988; MacKenzie 1988; Pickup 1988) or the regendering of urban space through specific mobility patterns (McDowell, 1999).

Although rigidly defined boundaries of public and private spheres have been replaced by less distinct boundaries, the factors that underpin this indistinctness and its fluidity are useful in exploring construction of gender, risk and mobility. Part of the public-private divide is predicated on risk experience, and the ascription of public space as fearful, an ideology that is produced and reproduced in patriarchy (Domosh and Seager, 2001; Lupton, 1999; Pain, 1997; Valentine, 1989; Weaver, Carter and Stanko, 2000). Within this discourse, women develop risk aversion as a result of notions of their physical vulnerability to men and fear of sexual violence. This negative experience of risk then impacts on the experience of other risks (Lupton, 1999) and contributes to a risk aversion based on public/private space that is passed through generations (Pain, 1994).

However, fear of space is just one aspect of risk experience and perception and therefore only partially explains women's interaction with the urban environment (Tulloch, 2000). Tulloch argues that risk experience is determined by a range of factors experienced over a long period of time and is dependent on spatiality and social contexts. Studies had already shown the differential risk experience associated with low income (Kelly et al., 1997), where low-income parents were more likely to recognise risks in both public and private space, as they lacked the resources to reduce risks inside the home. 'Better-off' parents externalised risks using semi-public space and coping strategies to control risk, contributing to increased automobilisation (Dowling, 1999).

It is the indeterminate space of the car that Sheller and Urry (2003) argue is making the public-private dichotomy irrelevant. Instead, they argue that this 'quasi-public' has become the context where gender roles are played out. They argue that public and private are relational and fluid and therefore there are no boundaries to be blurred. For them the 'distinction between public and private domains should be dispensed with since nothing much of contemporary life remains on one side or other of the divide' (Sheller and Urry, 2003, 122). However, it could be argued that the fluidity of these concepts, rather than making them obsolete, is leading to an iterative process of redefinition.

Whether through institutionalised ideologies of mothering or more localised everyday discourses, different cultures of mothering are established and re-established at different times and in different spaces. The penalties for not succumbing to the dominant cultures of mothering are manifold as they involve potential harm to children, as defined by 'experts', including state sanctions and localised cultures of blame. It could be argued, then, that risk underpins the very nature of parenting and contributes to constructing the parenting culture. Space and in particular mobility in space are inextricably linked to risk and motherhood. Motherhood hereby constructs space as space constructs motherhood, a concept that can be explained further through discourses surrounding the public-private space dichotomy. It is this relationship between gender and space that, together with cultures of risk and mothering, determine mobility. However, it is not only through global and local ideologies of motherhood, but also through mothering practices that mobility, risk and motherhood are co-constructed.

Multidimensionality and increased mobility

The norms and practices of mothering are also fluid and mobile and have changed considerably over the last century. As an increasing number of mothers enter the workplace (Office for National Statistics, 2004), with parallel increases in nursery and childcare places,[7] motherhood becomes defined by its multidimensionality and therefore its mobility. Gender constraints continue to have impacts. The workplace continues to represent a hostile environment for mothers, despite increased flexibility in working practices, and the demand for childcare continues to exceed supply (Skinner, 2005). In addition, women spend more time on childcare and household responsibilities than men, even if they are in full-time employment (Office for National Statistics, 2004).

These gender constraints are dependent on both the underlying ideologies that define cultures of motherhood and the ability of the space to accommodate changing patterns of mobility. Adequate infrastructure is therefore important, as 'transport also

7 There has been a quadrupling of nursery places in the UK since the mid-1990s (Skinner, 2005).

plays a key role in women's efforts to manage the multiple roles they play' (Hamilton and Jenkins, 2000, 1799). However, it is cultures of mothering that are bound up in everyday risk experience that seem to determine, as they are determined by, everyday mobility practices and decision making. This is demonstrated in Dowling's (1999) study of the interdependency of cultures of automobility and mothering in suburban Sydney. Dowling argues that everyday discourse is set within the context of dominant ideologies of mothering, which 'become points for negotiation in everyday practices of mothering, and to this extent are indirectly shared' (Dowling, 1999, 347). For Dowling, cultures of mothering are about how women relate to their children and their general approach to child rearing.

The constraints imposed by childcare responsibilities (Paull and Taylor, 2002) become constraints imposed by an increasingly mobile society, which increasingly encompass a mobility management role, co-ordinating children's movements throughout the day (Skinner, 2005). Skinner uses the concept of 'co-ordination points' to explore periods in the day when management of children's mobility is critical, finding that a significant number of such periods were managed by mothers within the household. She found that informal childcare was being used to transport children at key times between different kinds of childcare and educational settings. The mothers in Skinner's study therefore were not only involved in directly escorting their children between their various destinations, but also in imagined mobilities, both in organising additional trips for their children and in virtually taking these trips with them at the critical times of the day. Motherhood practices are not only increasingly mobile but potentially hypermobile, as risk experience coupled with the expectations of 'good' mothering result in mobility that is unsustainable. Mobility can therefore represent significant constraints on mothers' lives, just as gender roles can represent barriers to mobility.

Mobility constraints and freedoms

It has been argued (Hamilton and Jenkins, 2000; Pickup, 1988; Tivers, 1988) that mobility along with access to childcare are the main constraining impacts on women's activity patterns. As discussed, gender constraints operate both ideologically and in practice and are both spatial and embodied. Tivers (1988, 86) argues that 'the complexity of the gender role constraint seems much greater than the simple idea of having to be at certain places at certain times in order to attend the needs of children'. She maintains that the gender role constraint is the 'outward face of the ideology that assigns child-caring responsibility to women'. Law (1999) argues that mobility constraints are often embodied as movement is embodied with mothers often forced to move through space encumbered with children, thus redefining their body's physical boundaries. However it could also be argued that it is the social and physical structure of the urban environment that is restrictive. Spatial embodiment can, however, define both the nature of mobility and its purpose, with the urban environment reflecting the body that is 'normal' and 'risk-free'.

Constraints are also spatial as the gender-ascribed roles of mothers necessitate and impose a pattern of mobility that reflects their responsibilities within a patriarchal society (Massey, 1994). Mothers therefore exhibit complex travel behaviour characterised by multi-purpose trips that often demand travel beyond the capacity of the public transport system.[8] At the same time, mobility has been constructed, in the main, as a masculine concept. As Law argues 'men are expected to move between spheres, while women's mobility may be interpreted as transgression' (Law, 1999, 580). Women's immobility, due to barriers specific to them, can lead to their transport exclusion, which is exacerbated by the gender roles of motherhood. Paradoxically, one of the most masculinised forms of transport (Domosh and Seager, 2001), the private car, offers the potential to liberate women from mobility constraints.

Although studies have shown that women tend to use some forms of public transport, such as buses (London Research Centre, 1998; Transport for London, 2004) more than men, they also account, almost entirely, for the growth in private car use. However, many women, those on low incomes in particular, have limited access to motorised transport of any kind. In addition, mothers' everyday experiences of walking as explored by Bostock (2001) can exacerbate mobility constraints. In contrast to the automobile representing emotional investment in children (Sheller, 2004), Bostock argues that walking can be emotionally draining if it is the only option available. Often compounded by poor physical environments, the young mothers in Bostock's study encountered a number of problems while walking. These included fatigue and stress, psycho-social effects of looking after fatigued children, and restrictions to limited geographical areas lacking shops, services, and social resources. Access to public transport for all the mothers was limited due to high fares and the prioritisation of other resource demands. As Bostock (2001, 16) found: 'mothers used their bodies as a means to bridge the gap between responsibilities and resources'.

Discussion of the socially excluding impacts of being without access to a car is often underplayed in social studies of mobility. However, low income can be an important contributory factor in restricting women's mobility options (Pickup, 1988; Hamilton and Jenkins, 1989). Mobility, like risk, can constrain or liberate. Mothers' mobility is constrained by spatial and temporal factors as mothers adopt a multidimensional role. However, mobility, as well as reinforcing gender constraints through a failure to provide access, can also potentially liberate women from their gender roles. Although specifically referring to migration, McDowell (1999, 168) states: 'Travel, even the idea of travelling, challenges the spatial association between home and women that has been so important in structuring the social construction of femininity in the 'West'.'

Therefore, mobility can represent emancipation from the imposed cultures and practices of motherhood as well as being a constraint. In turn, motherhood can represent freedom in terms of identity and spatial security, as discussed. The sense of identity through motherhood can be a valued and positive element of women's lives. As Valentine (1997, 49) states: 'Although taking prime responsibility for children is a major constraint on women's lives which can undermine their own

8 In the UK women make more education escort trips than men; between the age of 30 and 39, 15 per cent of trips are 'escort to school'. Women made over 40 per cent more escort trips than men in 1999–2001 (Office for National Statistics, 2004).

sense of self, motherhood can also give women a sense of meaning and identity.' The sense of self that parenthood provides, therefore, can liberate women's lives as well as constrain them, both of which occur in the context of localised identities and cultures of mothering. There is also evidence that women establish support networks with other mothers after childbirth, which is important in locating and understanding their own experience (Tietjen, 1985; Urwin, 1985; Holloway, 1998). This period of motherhood can therefore represent a relatively mobile period, with a sense of freedom and autonomy gained through networking.

As increased mobility can be liberating, so too can increased automobility, most apparent when compared with the relative impacts of carlessness described by Bostock (2001), as discussed. Cars represent freedom from gender and generational constraints (Hamilton and Jenkins, 2000; Dowling, 1999). Their 'semi-private' space can represent a relatively 'safe' environment, a way of coping with risk as discussed. They can represent a more positive experience in escorting children (Bostock, 2001). Cars can go where public transport cannot, allowing multi-purpose trips that are often unfeasible by other means. They can facilitate emotional investment in children (Sheller, 2004) and 'facilitate 'family time'' (Dowling, 1999, 351) as some parents' cars can be a good space for one-on-one social contact with children (Barker, 2003). However, as Tivers (1988, 87) states: 'access to a car … may widen activity patterns without in any way changing the dominance of the overriding constraints'.

The car can represent freedom as a strategy for coping with risk in public space as well as a way of coping with mobility constraints based on mothering practices. However, both are more applicable to middle-class mothers. Mobility and risk are intertwined, with the nature of their co-construction based on a number of social and cultural factors. Where mobility is considered risky, studies have shown that immobility can be considered an effective control of risk. Domosh and Seager (2001) discuss immobility in terms of its challenge to gender roles. They cite the work of feminist geographers who show that immobility is not necessarily disadvantageous, that 'networks based on 'localness' and 'fixedness' can and do provide deep reservoirs of resources, reciprocity, assistance and comfort…' (Domosh and Seager, 2001, 121).

Conclusion

This chapter has discussed how motherhood, risk and mobility are based on dominant ideologies and localised cultures and are dependent on a range of socio-cultural and spatial factors. The co-construction of each of the concepts has been evidenced throughout the chapter. It has been argued that motherhood is a multidimensional concept encompassing the diversity of women and the impact of this diversity on global and local ideologies of mothering. Both global and local ideologies contain the constructions and reconstructions of what it is to be a 'good' or 'bad' mother, which are associated with cultures of blame associated with risk minimisation and control. It is also apparent that social status is a determining factor in terms of ideology and practices of motherhood, risk and mobility. Marginalised women are often constructed as 'bad' mothers, which in turn leads to further marginalisation.

The ambivalence that results from ever-changing constructions of good mothering facilitates the existence of a blame culture, where risky behaviour outside of established norms is unacceptable.

The culture of blame in mothering extends to the imposition of risky mobility on their children. This in turn acts as a mobility constraint for mothers. Although mobility in itself can be deemed as risky within dominant and local cultures of risk and mothering, there are apparent exceptions that are dependent on socio-cultural contexts. Within some white middle class cultures the car is a coping strategy, a way of overcoming risk and a symbol of good mothering. For others the only way to avoid risk is to remain relatively immobile, thus complying with local cultures of mothering predicated on risk aversion. There can be little doubt that mothering, along with other aspects of gender, is definitively shaped by mobility and its riskiness and in turn that changing intensities of mobility are determining what it is to be a mother.

References

Adams, J. (1995), *Risk* (London: UCL Press).

Adams, J. (1999), *The social implications of hypermobility* (Paris: OECD).

Adams, J. (2005), 'Hypermobility: A challenge to governance' in C. Lyall, and J. Tait, (eds.) *New Modes of Governance: Developing an Integrated Policy Approach to Science, Technology, Risk and the Environment* (Aldershot: Ashgate).

Backett-Milburn, K. and Harden, J. (2004), 'How children and their families construct and negotiate risk, safety and danger', *Childhood* 11 (4), 429–47.

Barker, J. (2003), 'Passengers or Political Actors? Children's Participation in Transport Policy and the Micro Political Geographies of the Family', *Space and Polity* 7 (2), 135–51.

Beck, U. (1992), *Risk society: towards a new modernity* (London: Sage Publications).

Beck, U. and Beck-Gernsheim, E. (1995), *The normal chaos of love* (Cambridge: Policy Press).

Bostock, L. (2001), 'Pathways of disadvantage? Walking as a mode of transport among low-income mothers', *Health and Social Care in the Community* 9 (1), 11–18.

Chase, S.E. and Rogers, M.F. (2001), *Mothers and children: feminist analyses and personal narratives* (New Brunswick: Rutgers University Press).

Clark, L. (1996), 'Demographic change and the family situation of children', in J. Brannen and M. O'Brien (eds), *Children in families* (London: The Falmer Press).

Connolly, D. (2000), 'Mythical dichotomies of good and evil. Homeless mothers in the United States', in H. Ragoné and F.W. Twine (eds), *Ideologies and technologies of motherhood: race class sexuality, nationalism* (New York: Routledge).

Dobbs, L. (2005), 'Wedded to the car: women, employment and the importance of private transport', *Transport Policy* 12, 266–78.

Domosh, M. and Seager, J. (2001), *Putting women in place* (New York: The Guilford Press).

Douglas, M. and Wildavsky, A. (1983), *Risk and culture* (California: University of California Press).

Douglas, M. (1986), *Risk acceptability according to the social sciences* (London: Routledge and Kegan Paul).

Dowling, R. (1999), 'Cultures of mothering and car use in suburban Sydney: a preliminary investigation', *Geoforum* 31, 345–53.

Duncan, S. and Edwards, R. (1999), *Lone mothers, paid work and gendered moral rationalities* (Basingstoke and London: Macmillan Press).

Dyck, I. (1990), 'Space, time and renegotiating motherhood: an exploration of the domestic workplace', *Environment and Planning D* 8 (4), 459–83.

Dyck, I. (1996), 'Mother or worker? Women's support networks, local knowledge and informal childcare strategies', in K. England (ed.), *Who will mind the baby? Geographies of child care and working mothers* (London: Routledge).

Fotel, T. and Thomsen, T.U. (2004), 'The surveillance of children's mobility', *Surveillance and Society* 1 (4), 535–54.

Furedi, F. (2001), *Paranoid parenting* (London: The Penguin Press).

Furedi, F. (2002), *Culture of fear* (London: Continuum).

Gillies, V. (2006), 'Working class mothers and school life: exploring the role of emotional capital', *Gender and education* 18 (3), 281–93.

Greater London Council (1985), *Women on the move* (London: Greater London Council).

Hamilton, K. and Jenkins, L. (1989), 'Why women and travel?', in M. Grieco, L. Pickup, and R. Whipp (eds), *Gender, transport and employment; the impact of travel constraints* (Aldershot: Avebury).

Hamilton, K. and Jenkins, L. (2000), 'A gender audit for public transport: a new policy tool in the tackling of social exclusion', *Urban Studies* 37 (10), 1793–800.

Hendrick, H. (1994), *Child welfare England 1892-1989* (London: Routledge).

Hendrick, H. (2003), *Child welfare: historical dimensions, contemporary debate* (Bristol: The Policy Press).

Hillman, M. (1976), *Transport realities and planning policy: studies of friction and freedom in daily travel* (London: Political and Economic Planning).

Hillman, M. (ed.) (1993), *Children, transport and the quality of life* (London: Policy Studies Institute).

Hillman, M. (1999), *The impact of transport policy on children's development*, Paper presented at the Canterbury safe routes to school seminar, Canterbury Christ Church University College.

Hillman, M. and Adams, J. et al. (1990), *One false move ... A study of children's independent mobility* (London: Policy Studies Institute).

Holloway, S. (1998), 'Local childcare cultures: moral geographies of mothering and the social organisation of pre-school education', *Gender, place and culture* 5 (1), 29–53.

Holloway, S. and Valentine, G. (2000), 'Children's geographies and the new social studies of childhood', in S. Holloway and G. Valentine (eds.), *Children's geographies: playing, living, learning* (London: Routledge).
Jackson, S. and Scott, S. (1999), 'Risk anxiety and the social construction of childhood', in D. Lupton (ed.), *Risk and sociocultural theory, new directions and perspectives* (Cambridge: Cambridge University Press).
Jenks, C. (1996), 'The postmodern child', in J. Brannen and M. O'Brien (eds), *Children in families* (London: The Falmer Press).
Joshi, M.S. and MacLean M. (1995), 'Parental attitudes to children's journeys to school', *World Transport Policy and Practice* 1: 29–36.
Kelley, P., Mayall, B. and Hood, S. (1997), 'Children's accounts of risk', *Childhood* 4, 305–24.
Law, R. (1999), 'Beyond 'women and transport': towards new geographies of gender and daily mobility', *Progress in Human Geography* 23 (4), 567–88.
Lewis, J. (1980), *The politics of motherhood* (London: Croom Helm Limited).
Lewis, J. (2005), 'Perceptions of risk in intimate relations at entry to partnership and with the arrival of children: the implications for social provision', *Social Contexts and Responses to Risk Launch Conference* (University of Kent, Canterbury).
Little, J., Peake, L. and Richardson, P. (1988), *Women in Cities* (Basingstoke: Macmillan Education).
London Research Centre (1998), *Women's travel in London* (London: London Research Centre).
Lupton, D. (1999), *Risk* (London: Routledge).
MacKenzie, S. (1988), 'Balancing our space and time: the impact of women's organisation in the British city 1920-1980', in J. Little, L. Peake and P. Richardson (eds), *Women in Cities* (Basingstoke: Macmillan Education).
Marshall, H. (1991), 'The social construction of motherhood: an analysis of childcare and parenting manuals', in A. Phoenix, A. Woolett and E. Lloyd (eds), *Motherhood* (London: Sage).
Massey, D. (1994), *Space, Place and Gender* (Cambridge: Polity Press).
McDowell, L. (1999), *Gender, Identity and Place: understanding feminist geographies* (Cambridge: Polity Press).
Miller, T. (2005), *Making sense of motherhood* (Cambridge: Cambridge University Press).
Mythen, G. (2004), *Ulrich Beck, a critical introduction to the risk society* (London: Pluto Press).
Murray, L. (forthcoming), *Mobilities and the school journey: mothers, children and the negotiation of risk landscapes*, Unpublished PhD thesis.
Office for National Statistics (2004), *Focus on gender* (London: Office for National Statistics).
Pain, R. (1994), *Kid gloves: children's geographies and the impact of violent crime* (Newcastle: University of Northumbria).
Pain, R. (1997), 'Social geographies of women's fear of crime', *Transactions of the British Geographical Society*, 22, 231–44.
Pain, R. (2003), 'Youth, age and the representation of fear', *Capital and Class*, 60, 151–71.

Paull, G. and Taylor, J. (2002), *Mothers' employment and childcare use in Britain* (London: Institute for Fiscal Studies).

Phoenix, A. and Woolett, A. (1991), 'Motherhood: Social construction, politics and psychology', in A. Phoenix, A. Woolett and E. Lloyd (eds), *Motherhood, meanings, practices and ideologies* (London: Sage).

Pickup, L. (1988), 'Hard to get around: a study of women's travel mobility', in J. Little, L. Peake and P. Richardson (eds), *Women in Cities* (Basingstoke: Macmillan Education).

Rose, G., Kinnaird, V., Morris, M. and Nash, C. (1997), 'Feminist Geographies of environment, nature and landscape', in Women and Geography Study Group (ed.), *Feminist Geographies, Explorations in Diversity and Difference* (Harlow: Longman).

Rosenbloom, S. (1992), 'Why working families need a car', in M. Wachs and M. Crawford (eds), *The Car and the City: The Automobile, The Built Environment, and Daily Urban Life* (Ann Arbor: The University of Michigan Press).

Scott, A. (2000), 'Risk society or angst society', in B. Adam, U. Beck and J. Van Loon (eds), *Risk Society and Beyond* (London: Sage).

Sheller, M. (2004), 'Automotive emotions', *Theory, culture and society* 21 (4/5), 221–42.

Sheller, M. and Urry, J. (2003), 'Mobile transformations of 'public' and 'private' life', *Theory, culture and society* 20 (3), 107–25.

Sheller, M. and Urry, J. (2006), 'The new mobilities paradigm', *Environment and planning A* 38 (2), 207-26.

Silva, E.B. and Smart, C. (1999), 'The 'new' practices and politics of family life', in E.B. Silva and C. Smart (eds), *The new family?* (London: Sage Publications).

Skinner, C. (2005), 'Coordination points: a hidden factor in reconciling work and family life', *Journal of Social Policy* 34 (1), 99–119.

Slovic, P. (2000), *The perception of risk* (London: Earthscan publishing).

Tietjen, A.M. (1985), 'The social networks and social support of married and single mothers in Sweden', *Journal of Marriage and Family* 47, 489–96.

Tivers, J. (1988), 'Women with young children: constraints on activities in the urban environment', in J. Little, L. Peake and P. Richardson (eds), *Women in Cities* (Basingstoke: Macmillan Education).

Transport for London (2004), *Expanding Horizons: Transport for London's Women's Action Plan 2004* (London: Transport for London).

Tulloch, J. (1999), 'Fear of crime and the media', in D. Lupton (ed.), *Risk and Sociocultural theory* (Cambridge: Cambridge University Press).

Tulloch, J. (2000), 'Landscapes of fear', in S. Allan, B. Adam and C. Carter (eds), *Environmental risks and the media* (London: Routledge).

Tulloch, J. and Lupton, D. (2003), *Risk and everyday life* (London: Sage).

Urry, J. (2000), *Sociology beyond societies, mobilities for the twenty-first century* (London: Routledge).

Urwin, C. (1985), 'Constructing motherhood: the persuasion of normal development', in C. Steedman, C. Urwin and V. Walderdine (eds), *Language, gender and childhood* (London: Routledge).

Valentine, G. (1989), 'The geography of women's fear', *Area* 21, 385–90.

Valentine, G. (1997), 'My son's a bit dizzy. My wife's a bit soft: gender, children and cultures of parenting', *Gender, Place and Culture* 4, 37–62.
Valentine, G. (2004), *Public space and the culture of childhood* (Aldershot: Ashgate).
Weaver, C.K., Carter, C. and Stanko, E. (2000), 'The female body at risk: media, sexual violence and the gendering of public environments', in S. Allan, B. Adam and C. Carter (eds), *Environmental risks and the media* (London: Routledge).
Wyness, M. (2000), *Contesting childhood* (London: Falmer Press).
Wynne, B. (1998), 'May the sheep safely graze? A reflexive view of the expert-lay knowledge divide', in R.E. Lofstedt and L. Frewer (eds), *The Earthscan reader in risk and modern society* (London: Earthscan).

Chapter 5

'Mobile Belonging': Exploring Transnational Feminist Theory and Online Connectivity

Michaela Fay

Introduction

In this chapter, I propose two things. First, that, if we are to understand contemporary mobility, we need to consider both geographical movement of subjects as well as the social, cultural, political, and, crucially, theoretical mobility within which movement takes place. Second, I argue that, in the light of recent trends toward the internationalisation of academia and the precarious position academic feminism takes within this trend, the idea of 'mobile theories' is of particular importance in the context of transnational feminist (online) networks.

My point of departure is Hannam et al.'s (2006) suggestion that there has been a 'mobility turn' in the social sciences and that we can thus be said to be faced with a 'new mobilities' paradigm'. That is to say, mobility needs to be reconsidered as a multi-layered concept, rather than the mere accumulation of miles travelled. Within feminist theory, Rosi Braidotti's (Braidotti, 1994) metaphor of the nomad is an example of such a position. Empirically, this chapter rests on my cyber-ethnographic research (online questionnaires, analysis of mailing list and website content, but also my own participation) of the International Women's University 'Technology and Culture' (*ifu*) 2000 and its participants. The empirical examples I offer throughout this chapter focus on some of my research findings, online questionnaires and email exchanges in particular, in order to explore how the members of this particular feminist network 'make sense' of the interplay between theory, feminism, and mobility. The chapter revolves around and departs from a concrete localised example – *ifu*. However, in the light of the overall current trends of European academia towards interdisciplinarity, internationality, but also marketability of the academy (of which internationality is a part) we can consider *ifu* to be indicative of these trends. Parallels can be drawn to other European feminist networks.[1]

1 For example the Athena Network (http://www.let.uu.nl/womens_studies/international/athena.php), the NOISE European Network of Women's Studies http://www.let.uu.nl/womens_studies/international/noise.php) or the newly launched WeAVE Network (http://www.weave-network.eu).

Context: The International Women's University 'Technology and Culture' 2000 (*ifu*)

During the summer of 2000, a number of universities across Germany hosted the International Women's University 'Technology and Culture' 2000 (*ifu*). Inspired by the slogan '100 days for 100 years',[2] *ifu* was designed as a three-month postgraduate programme with a decidedly international, women-only student body; an interdisciplinary curriculum; and a theoretical focus on women and gender. Overall, the project consisted of a number of components and groups of actors. The curriculum was organised into 6 'Project Areas' (PAs) – Body, City, Information, Migration, Water, and Work – and individuals applied to participate in one of these.[3] In total 747 women from 105 countries participated in the *ifu* academic programme. In addition, 313 lecturers from 49 countries contributed to *ifu*. Alongside these two main groups of academic actors, *ifu* was equipped with a large number of administrative staff, as well as 74 academic tutors. In addition, the project comprised three further components, a Service Centre,[4] an Open Space programme[5] and lastly, a virtual component – Virtual International Women's University, v*ifu*. The women who attended the event as participants were from a vast array of walks of life. Mainly, they had an academic background (predominantly, but not exclusively, in the social sciences, including Gender and Women's Studies). A substantial number of participants had a professional or vocational background (for example in law, teaching, NGO work, human rights, medicine, developmental aid work, social work). Further, *ifu* specifically invited artists to participate in the event, in order to put into practice what was designed as an 'art and science dialogue'. The latter was intended to contribute to *ifu*'s aim to challenge traditional ways of knowledge production.

Ifu was designed both as emerging from and contributing to the reforms in higher education in Germany in particular and in Europe more generally.[6] Among these are developments toward increased internationality of student bodies, including a rise in student mobility; the reorientation of academic disciplines toward interdisciplinary research, as well as the trend toward the corporatisation of universities. When the idea of *ifu* materialised, its creators had hopeful and utopian visions in mind about what such a space would be like, who would inhabit it and how it would change

2 The reference is to 100 years of women's movement.

3 I myself attended the event as a participant in Project Area Body.

4 The Service Centre was designed as a 'one stop', centralised administrative structure, consisting of front offices and back offices which were designed to be in constant communication with each other. From the participants' point of view, the Service Centre equally provided help with administration, registration, housing, as well as more mundane concerns such as everyday life in Germany and doctor's visits (see Metz-Göckel, 2002, 259).

5 The Open Space programme was the only component of *ifu* accessible to the public. Based in a popular cultural forum in Hannover, one day a week, the Open Space programme offered a fair-like array of lectures, film screenings, food, exhibitions and entertainment for *ifu* participants and members of the public alike.

6 As emerging from and determined by the Bologna process, the implementation of which began with the 1999 Bologna Declaration.

the world. As a women-only space it should set new examples for the academy as a whole. In particular, it should:

- Address and represent trends toward student mobility by recruiting participants from across the world.
- Accommodate an international group of participants by conducting the event in English.
- Provide sufficient funding to enable a 'needs blind' admission system, by which participation should not be determined by economic factors but rather by academic ability.
- Offer a dense, high quality, interdisciplinary curriculum with a focus on gender, revolving around the awareness of how gender, race, and class deeply structure all aspects of life.
- Not least, offer its participants the opportunity to form deep connections not only with each other but also with the institution itself in order to emulate a graduate network,[7] similar to those of American ivy-league colleges.[8]
- Such networking should be sustained by having a virtual element functioning as a platform for enduring connectivity among participants. The virtual platform of such a university should bring *ifu* en par with trends toward e-learning and networking and should come with adequate training to enable those with less or inadequate access to, and knowledge of, the use of communication technologies to 'catch up' with those whose access to these technologies is less obstructed (see, for example, Schelhowe, 2001; Gürses, 2001).

Academic life in particular is one area that is characterised by a sometimes overwhelming onscreen-ness. It is thus not surprising that *ifu*'s virtuality has been regarded as representative of 'the development and implementation of visions of the 21st century university' (Müskens and Hanft in Metz-Göckel, 2002, 346, my translation). Indeed, *ifu*'s long-term existence in cyberspace was not considered the 'next best thing to the real thing' but a necessary and timely extension of and addition to *ifu*'s onsite-ness. Indeed, the ongoing online activity and exchange among *ifu* participants five years on demonstrate the 'realness' of v*ifu*.

A 'virtual university', as Heidi Schelhowe (2001), coordinator of the v*ifu* project (at the time based at the Institute of Computer Science at Humboldt University in Berlin) explains, means initially nothing other than a traditional university set-up facilitated by computer technology to allow for long-distance learning. That was, however, *not* what the virtual component of *ifu*, v*ifu*, wanted to achieve. Rather, v*ifu* was imagined as a 'lived-in' space, a user-friendly platform for networking in which the focus was on the 'being together', rather than technological know-how. So when *ifu* became v*ifu*, a corporeal, on-site community was meant to become a virtual one – a networked web of attachments and connections. This network, it was hoped, would further aid 'relating the global and the local'. Virtual connections were not so much seen as being created by electronic networks in and of themselves, but rather as an

7 The female equivalent of the idea of the 'old boys' network'.

8 See Ayla Neusel (2000).

'intercultural exchange' in which one's local context becomes visible and forms the 'basis for mutual understanding and diversity at the same time'. In sum, v*ifu* enabled communication between *ifu* participants beyond *ifu*'s three-month on-site period, at a time 'when you have to speak and act in local contexts with the new experiences and the knowledge you have gained at *ifu*' (Schelhowe, 2001).

The networking platform of v*ifu* is a straightforward virtual space, essentially no more than a website including a number of notice boards and mailing lists. When you visit www.v*ifu*.de, this is what you find:

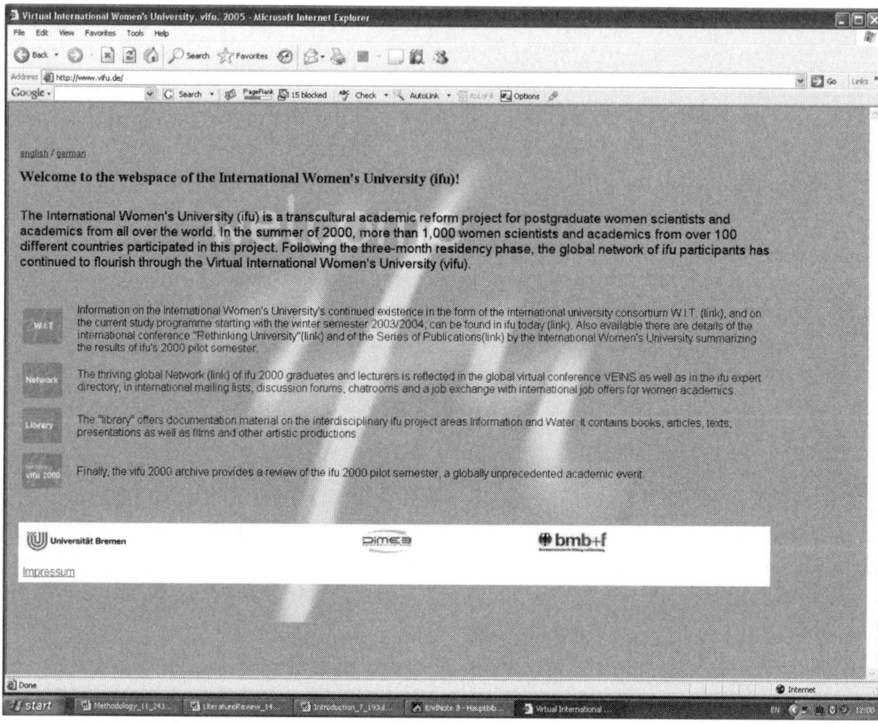

Ifu's aims, as I have laid them out above, are ambitious. Through them, the event is presented as hopeful for a better future for academic feminism, and determined to enhance women's opportunities to network across disciplinary and geographical borders. The event's aims position it somewhere uncomfortably between mainstream and innovation, exclusivity and marketability, local accountability and global dreams. It is my suggestion that *ifu*'s aims, as listed above, raise a number of crucial questions about feminism, mobility and changes in the nature of the academy.

What kind of feminism, for example, does an event promote and enable that is simultaneously trying to transcend traditional structures and be accountable to them? This is a tension academic feminism in particular has felt since the beginning of its institutionalisation (Skeggs, 1995). Moreover, given *ifu*'s internationality as represented by its participants' mobility and diversity, it is necessary to ask how belonging – to each other as well as to the event – is negotiated within these tensions. And crucially, we must ask how belonging and mobility, both in the real space of the event and in cyberspace of v*ifu*, are realised and experienced.

In response to such questions, I consider *ifu* itself as an event that is located within the discourse of mobility and transnational feminism. I juxtapose my critical analysis of it with an exploration of how the event's participants negotiate and experience belonging within the framework of mobility and transnational feminism. I consider these questions through this particular event because, as Sabina Mihelj (2002) suggests, *ifu* can be seen as a condensation of the internal struggles of feminism. Furthermore, for all its particularities, it reflects a crucial point in the history of feminism, feminist academia, and feminist theory, in particular the trend toward an increase in mobility – both intellectual and geographical. Conceptually, since the rise of standpoint theories, feminism has been concerned with questions of location and situatedness (Haraway, 1991; Hartsock, 1998; Rich, 1984). That is to say, in feminist theory, arguably more so than in other fields of inquiry, 'where one comes from' matters with regards to the (power) negotiation of race, class, sexuality, geography. As an academic discipline, feminism (and Women's Studies more particularly) relies in the 'migration' of scholars from other disciplines into Women's Studies. Often this means changing academic frameworks as well as institutions and even countries. One of the challenges of academic feminism, as Naomi Schor and Elizabeth Weed have put it, is thus to avoid 'sticky monotony', to 'remain constantly on the move, to keep buzzing. And it is striking, in fact, that [much of feminist research] calls for positionality and a politics of location [which] go hand in hand with an emphasis on mobility' (1990, preface).

Through *ifu* we can thus begin to explore the complexities of what it means in these times to make global feminist connections. In a context in which mobility shapes subjectivities as much as it does in the population of *ifu* participants, the interplay between intellectual work and the making of home is crucial for the making of such connections. One of my research participants summarised it most compellingly:

> C and I are still working on our presentation of what *ifu* was/is all about [...] We called it 'Academia Meets Home: The International Women's University, Hannover 2000', because we really feel that our experience of *ifu* this summer was, in the sense, a place where our scholarly studies intersected with making a home. And that's rare in an institutional setting. That's why we think that *ifu* really worked, despite all of the chaos, the crunched intensity, the gaps, and the problems. *Ifu* actually became a home. (Posted on 23/10/2000)

Mobility/ies

It is safe to say that throughout the past decade, mobility as an analytical tool has become omnipresent. In this light, Hannam, Sheller and Urry argue for the emergence of a 'new mobilities' paradigm' (2006, 2) which begs us to account for 'mobilities in the fullest sense' thus challenging social science to change both the objects of its inquiries and the methodologies for research. As the materiality of mobilities becomes more complex, Hannam, Sheller and Urry argue, we need to be asking different questions about the conditions of mobility. This includes the premise that mobility cannot be described without taking into account the 'spatial, infrastructural and institutional moorings that configure and enable mobilities' (2006, 3), nor can it be described without acknowledging what Doreen Massey has called the 'power geometries of everyday life' (Massey, 1993). Arguably this is nowhere more obvious and important than in relation to academic feminism and transnational feminist networks.

If we take *ifu* to be part of a world that is imagined as hypermobile and, more specifically, if we take it to be part of the imagining of a network of mobile feminist academics, it is obvious that the paradigm of mobility is pivotal. However, we should not forget that some individuals move through the world more easily, safely, and often with more agency than others. Also, and important in the context of this chapter, mobility is gendered in more than one way (Brah, 1996; Spivak, 1996; Yuval-Davis, 1997). Historically, as Nira Yuval-Davis has suggested, women have been constructed as the antithesis of mobility by being represented as the bearers and conservers of nations and their boundaries. And yet, Rosi Braidotti suggests that 'perhaps women are historically nomadic, in that we are not yet first-class citizens'. This, she argues, applies also to mobility as 'the intellectual space of creativity'. Women have not traditionally had the 'freedom to invent new ways of conducting our lives, new schemes of representation of ourselves. [...] And yet the challenge for the women of today is to conjugate the positive aspects of this nomadic condition with something that we would call responsibility for and accountability to our gender' (Braidotti, 1990, 118). It could be said that one purpose of academic feminism is precisely such coupling.

Women's agency in mobility has been changing, certainly over the past two decades. Forced migration and trafficking are no longer the immediate and only associations between women and movement.[9] The very emergence of mobility studies (in addition and often in contrast to migration studies) can be read as an outcome of such and other changing patterns in mobility. Where 'actual' mobility is concerned, women, as Braidotti and others suggest, have come a long way from their construction as immobile and stable per se (Braidotti, 1994). In fact, it is my aim here to contribute to the creation and understanding of women *as* mobile agents. The following illustration of the mobility patterns and understandings of *ifu* participants, who also participated in my research, illuminates this agency.

Many of the stories related to me throughout my fieldwork have an almost confessional air and, as one of the women in my research has told me, my questions tapped into what they themselves ponder every day – often under a considerable amount of pain. Responding to my research became for many women a welcome opportunity to reflect on their own contradictory relationship to their mobility. Most of my research participants are originally from Western countries, ranging from Britain, Sweden, Germany, Italy to Greece, and the Netherlands but going as far as India and the United States as well as Israel, Turkey, Croatia and the Ukraine. In three cases, the country of origin and nationality differ from each other. Almost none of

9 Although these have by no means disappeared and remain deeply gendered migratory practices.

them, however, were at the time of fieldwork resident in their country of origin – and surprisingly, all those women who were, were German, which makes this group of research participants the least mobile. Most members of this small group of women have also not actually lived in any other countries apart from Germany.[10] Current countries of residence span a similar field to countries of origin. They range from Canada to Estonia and include the UK, Germany, Sweden, Spain, Austria, France and the Netherlands. The length of residency in any given case differs widely and ranges from 'all my life' to just a few months, but averages approximately three-and-a-half years. Most of my research participants have lived in two or three different countries and most of them on a continent different from their original home place. There are a number of reasons for the women's mobility but in most cases, 'work' or 'research' is mentioned as the motor, and in one case 'father's profession'.[11] The legal status of research participants in their respective countries of residence is either as a citizen, a permanent resident or on a visa (mostly as a student). One of them is resident as a political refugee and one is classified as 'undocumented'. All of my research participants share a similar professional background, namely in research and teaching. Most women stated they live in a long-term romantic relationship and approximately half of them live with a partner of a different nationality from their own. This applies equally to those women who define themselves as 'straight' and those who define themselves as 'queer' (approximately two-thirds and one-third of participants respectively). Many of these relationships are long distance. Mobility is thus not completely restricted to the professional realm but is also, one could argue, of a 'cultural kind' in so far as there will be a number of negotiations (cultural as well as political and bureaucratic) that arise from these relationships. In some instances 'love' is mentioned as the very reason for one's mobility and these women have decided to move to where their partners were located (ideally combined with a professional move/development of their own). It is perhaps not surprising that these mobile relationships are childless. In fact, only one woman who participated in my research has children and her biography is one that is marked by the absence of geographical mobility.

Understandings of one's own mobility vary greatly among the women who participated in my research, although most women would define themselves either as global citizens or travellers. A number of women identified with the category of the nomad[12] and a small number thought of themselves as migrants, one woman

10 I remember being really surprised at the time of conducting fieldwork at the fact that these women replied to my research call in the first place, because my call looked explicitly for women who were mobile. However, as I learned later, the absence of 'actual' mobility did not mean that these women had not spent a good deal of time reflecting on mobility, including their own lack thereof, and what that might mean in relation to their awareness that this made them, in fact, 'different'.

11 This particular woman's father used to be the manager of an international hotel chain and mobility has been a key feature in her life from early childhood on. Clearly her biography is the one out of my research participants that is marked by the highest level of mobility.

12 Although one or two of them added 'not in the Braidotti sense'. When I pursued this further, I was told that they took nomadic to mean, more colloquially, that one stays in a place for a while and then moves on and so on. A nomadic lifestyle came to resemble something

respectively as refugee and as exile and two as diasporic. *Dis*identifications corresponded in a contrasting manner. That is to say, those women who identified their mobility as diasporic or exilic or thought of themselves as a refugee were most likely to have a strong disassociation with categories such as global citizen. And vice versa, women who thought of themselves as travellers tended to disidentify strongly with migrant or refugee. The latter, they explained, has strong connotations with force and displacement whereas they perceived their own mobility as more privileged. Other categories given were 'restless', 'migrant by choice', 'homeless', 'illegal', 'unsettled', 'foreigner' and 'globe trotter'. With respect to *ifu*'s aim to attract internationally mobile women, the group of participants who contributed to my research certainly matched this expectation. What became strikingly obvious among this group of women was the high level of everyday introspection and reflexivity these women engaged in with respect to their mobility. This quote by one participant exemplifies this: Here she describes the network of *ifu* participants as:

> A site for dislocated subjects, both literally and figuratively (that is, inbetween cultures, homes, sexualities, etc) [who] meet in the interstices of shifting and multiple locations…

When considering togetherness and belonging in this particular feminist network, I thus do not understand these two modes by the evidence of mobility practices but rather I ask how they are constituted and experienced within a context of global mobility.

Examining the relationship between language and mobility, for example, it can be argued that *ifu*'s globality is represented by the fact that, although in many ways deeply 'German', its communal language is English. Participant narratives illuminate some of the emotional and political consequences of life outside one's mother tongue. Based on online interviews which trace *ifu* participants' mobility and their changing relationship to their mother tongue as well as their experience with English as a 'global language' and *ifu* as a global event, I found there to be an emergent population of 'native second-language speakers' for whom English is indeed, as Sneja Gunew puts it, their 'most portable of accessories' (Gunew, 2003, 41), which is both foundational and expressive of mobility and internationality.

Mobility, it will have become apparent through the above claims and examples, cannot be divorced from a context of globality. Following Franklin, Lury and Stacey (2000) I take globality to be 'not simply an empirical force that has changed the everyday realities of people's lives but […] a discursive condition currently being reproduced within academia and outside it' (2000, 4). As a discursive condition, they understand investigating the global 'as a fantasy, as a set of practices, and as a context' (2000, 5). It is important to keep sight of the fact that the mobilities paradigm emphasises the ways in which the world is *perceived* as hyper-mobile. That is to say, mobilities studies begin with the premise that the world is largely imagined or dreamed of as 'hyper-mobile' and 'fluid'. 'Mobility', as Franklin et al. argue, is

---

of a homeless lifestyle, where one puts down 'just enough' roots to be able to make a home for the time being but also to easily uproot oneself again. 'The Braidotti sense' of nomadic was rejected on the basis of what they felt was 'over theorised' and thus useless as a tool for everyday life.

something that is put to work to produce desires, anxieties, fears, identities, and so on, in the imagining of the globalised world as a borderless world. When examining mobility it is thus by no means merely 'movement' that requires attention. Rather, what requires examination is, in particular, the recent diversification of the *kinds* of mobility – theoretical, political, social, affective – that shape the world. It is also the way in which we need to think about mobility itself – the locations and conditions of mobility and what we understand as mobility in the first place – that needs to be rethought. One such diversification, I suggest, is the increase in online dwelling that has come to shape much of our lives.

Cyberspace

The emergence of the Internet and communication technologies has been changing our understandings of place and belonging profoundly. When it was first emerging, 'cyberculture' (Escobar, 1994) was heralded as a transformation of the nature of social life and cultural identity. After an initial surge of excitement on the part of some as to the boundless possibilities for the emergence of online identities, theorists have come to take cyberspace and the ways in which individuals inhabit it as thoroughly embedded in a cultural, social, political and economic context (for example, Balsamo, 1996; Gray, 1995; Gajjala, 2001). The discussion in the literature has moved from the simple opposition of 'real' and 'virtual' to more nuanced accounts of the imaginaries of the former and the embodied materialities of the latter. Forms of connectivity and ways to connect, including questions of power and inequalities, are certainly at the heart of these explorations. One thread of research that has looked at the role of cyberspace argues that the world is simply becoming smaller due to the omnipresence of virtual dwelling. The idea of the world as a global village, however, conjures up perhaps misleading images of comfort, as well as a certain sense of stability. Others have examined cyberspace as a place in its own right and with its own possibilities/modalities of being together (Adams and Warf, 1997; Star, 1996). Where most scholars agree, despite these contrasting perceptions of cyberspace and globalisation, is that through the increased possibilities of technologically mediated communication, one's physical location is no longer the only reference point for belonging, nor for interaction. That is to say, one can be in any one place physically, while simultaneously dwelling elsewhere via the use of communication technology. As Roseanne Stone's argument suggests, computer mediated forms of togetherness are 'incontrovertibly social spaces in which people still meet face-to-face but under new conditions of both 'meet' and 'face'' (Stone, 1991, 85).

Furthermore, it has been debated whether and how the construction of not only social space but also identity has been influenced by the emergence of cyberspace. Some argue that cyberspace enables identities themselves to become flexible and mobile, allowing people to move in and out of what used to be fixed identities or alternatively to adopt multiple identities (Plant, 1997; Turkle, 1996). Such fluid approaches to identity in cyberspace have, on the one hand, been criticised for erasing inequalities of race and class (Nakamura, 2002) while, on the other, been heralded for enabling individuals precisely to playfully overcome these inequalities (Turkle, 1996; Danet, 1998).

Following on from arguments such as Stone's, (1991) there is further debate about whether or not virtual communities are indeed 'real' (Rheingold, 1993). For, if anyone could potentially be anyone online, it has been asked, where is the 'substance', the 'ground' of togetherness that is so often associated with 'real' communities? However, rather than thinking of real space in opposition to and separate from cyberspace, I want to suggest here that, in order to understand multiple expressions and experiences of mobility, it is more useful to think of them *in relation* to each other.

John Urry's work offers a useful step in this direction in his analysis of physical co-presence in *relation to* virtual dwelling (Urry, 2002; 2003). His argument allows us to think of the flows of people in connection with the flows of electronic messages, rather than thinking of each of them as confined to their disconnected respective spaces. Furthermore, Urry's argument invites us to think about what kinds of physical movement might be triggered by online interaction, rather than merely the other way around.

Virtuality, as indicated above, is an integral part of the event *ifu*, and onsite-*ifu* and online-*ifu* are intimately connected. It is important to analyse how they differ from each other but also constitute each other as places of attachment and detachment as well as the kinds of mobility they invite, allow for, or alternatively, hinder. Each space can thus be read as creating different conditions of possibility of belonging.

Looking at some of the ripple effects of the event's mobility and globality help to illuminate this proposition. When contemplating the relationship between this onsite event and its online extension, for example, one quickly detects the importance of the notions of home and belonging. After having dwelled at this international event, participants report finding themselves 'returning home with a difference' (Ahmed, 2000). The transition from *ifu* as an onsite event to the virtual platform v*ifu* demonstrates in particular how in their narratives of returning home, participants reconfigure not only the meaning of home but also construct *ifu* and v*ifu* respectively as places of belonging that enable specific kinds of attachment. 'I experience the sweet feeling of being home now' is how one participant puts it, 'but also it is hard leaving another sort of home behind – the *ifu* home' (posted on 25/10/2000). And more succinctly even, another participant describes the activities in this online space as 'mousekeeping'. This, she adds, is:

> not to be confused with housekeeping although [...] there are some connections once we think about the concept of home and where that actually exists... (Posted on 27/10/2000)

Likewise, participants' negotiations of the boundaries of the v*ifu* community demonstrate a complex engagement with mobility and togetherness. Belonging to the online space v*ifu*, it is argued in a debate surrounding the possible admission of a new member who had not previously attended the onsite event, requires the shared experience of having attended the onsite event *ifu*. Having-been-there, is thus constructed as a (arguable) foundational requirement for belonging to the online network. The experience of having been at the event is utilised by some participants and rejected by others as constitutive of a particular '*ifu* woman' and '*ifu* community' which is conceived as a node in the wider network of feminist websites and e-lists. These examples indicate how important it is to pay specific attention to the potentially increasing importance online connections will play for feminist togetherness and how these might transform the ways in which we think about belonging, and also mobility.

Mobile belonging

The existence of an event and network such as v/*ifu* can be said to echo and illuminate recent developments in feminist theory as well as in Women's Studies as an academic discipline (Brown, 1997; Martin, 1997; Scott, 1997; Skeggs, 1995; Wiegman, 1999/2000). It resonates with the engagement of feminist scholarship with increased mobility, most notably perhaps in the form of theories of transnational feminism (Kaplan and Grewal, 2002). It is located within the debate about the place of Women's Studies within the academy more generally. And more broadly, it reflects the changing nature of the academy as a whole, including an increase in both intellectual and geographical mobility. *Ifu* and its participants, in often contradictory ways, reflect and respond to all of these.

It is my suggestion that the link between feminist theory and mobility can be understood by exploring the dynamic relationship between physical, geographical movement and mobility as an intellectual and theoretical position.

Ifu itself, as well as the theoretical ideas that frame it, are located within the discourse of what Caren Kaplan and others have called transnational feminism (Alarcon et al., 1999; Brah, 1996; Mendoza, 2002). We ought to take seriously the ways in which feminist theories as well as the lived experiences of belonging to feminist networks have increasingly become mobile, moving from a politics of location and situatedness to a politics of affinity and networked coalitions, as Kaplan has argued in *Scattered Hegemonies* (Kaplan, 1994). A feminist politics of location, she suggests is 'not useful' when it merely reflects and reaffirms 'authentic, primordial identities' (1994, 139) and thus only 'makes room' and 'creates space' for those who are not already at the centre of the Western mainstream. As is common in her work, Kaplan carves out a strong link between postmodernism, feminism and transnationality. She points out that 'conventionally, 'global feminism' has stood for a kind of Western cultural imperialism' (1994, 17) and argues for an acknowledgement of 'transnational flows' (ibid.) both culturally and economically, in order to fully and appropriately understand the conditions that shape women's lives.

A transnational feminism (Kaplan, 1994) thus acknowledges (and even emerges from) mobility by taking the movement of people and theories as a premise from which to produce knowledge and form alliances. It is aware of the processes of globalisation and its inherent inequalities, thus remaining critical of the hailing of the 'anything goes anywhere' globalisation enthusiasts. An interdisciplinary production of knowledge can be said to rest on the basis of intellectual mobility in so far as it transgresses the boundaries of disciplines. And finally, it is a mobile feminism in so far as it acknowledges the situatedness of subjects along the lines of not only gender but also race, class, and sexuality and urges us to be reflexive about these and the ways in which they interact and have effects on one's location but do not determine the latter. Women's Studies students can be said to emerge as 'agents of change' (Morley, 1999), as 'compromisers, pragmatists and innovators' (Griffin and Hanmer, 2003, 15). This, they conclude, is 'partly fostered by Women's Studies' critical stance in relation to traditional forms of knowledge production and its interrogative position, partly because having to defend the subject and engage persistently with critiques both within and about the discipline fosters the ability to argue effectively' (Griffin and Hanmer, 2003, 13).

Rosi Braidotti's (1994) figuration of the nomad can be used as a point of departure that enables us to understand changing migratory practices as well as the links between mobility and subjectivity. Braidotti describes the nomad as a figurative tool, which she uses in order to understand a post-modern feminist subjectivity *as well as* mobility practices. Braidotti's argument suggests that mobility is never limited to movement. Rather, Braidotti's nomad (nomadic thinking) offers tools to work through 'established categories and levels of experience' and allows one to 'blur [...] boundaries without burning bridges' (1994, 4). Mobility beyond geographical movement can thus be contemplated as an 'epistemological position' (1994, 23). Women's Studies can be read as the manifestation of such a proposition. As Braidotti (2002), Skeggs (1995) and others (Hemmings, 2006; Mohanty, 2003; Wiegman, 2002) have suggested, Women's Studies' position within the academic context is shaped by mobility in the sense of being 'adaptable' and on the move both theoretically and within institutional structures. Similarly (or rather, extending these suggestions), Griffin and Hanmer (2003) propose that Women's Studies creates 'mobile subjects'. This proposition is pivotal to my argument, because it invites us to think about mobility in ways that include, but are not limited to, physical movement. Griffin and Hanmer's study[13] offers a number of findings about Women's Studies students and their academic trajectories, including intellectual as well as physical mobility. First, it demonstrates that taking Women's Studies is a *conscious life choice* for most women. One does not stumble into this particular discipline. Secondly, Griffin and Hamner describe the *transformative potential* of undertaking Women's Studies. That is to say, as one engages with the construction of gender and women's experiences, one begins to see oneself and one's place in society in a different light. Women's Studies becomes thus the looking glass into change and, often, political engagement. Griffin and Hanmer (2003) report two connections between studying Women's Studies and being physically mobile: the lack of Women's Studies in one's

---

13 The study was funded by the European Commission and is based on empirical research carried out in nine European countries (Finland, Germany, Hungary, Italy, France, Spain, Slovenia, the Netherlands, and the UK), during 2002. It combines non-random quantitative data elicited through questionnaires from an average of 50 past and 50 current Women's Studies students in each participating country with qualitative data derived from 30 semi-structured interviews (20 past and 10 current Women's Studies students per participating country). The latter were carried out in order to establish the meanings of some of the answers provided in the questionnaires, and to lend depth to the data gleaned from the questionnaires. The study is thus based on over 900 questionnaire returns and 270 interviews (Griffin and Hanmer, 2003, 3). The project and all the data reports are available at www.hull.ac.uk/ewsi.

home institution means that travel is necessary, or the existence of Women's Studies is discovered by chance during a study or work exchange. The latter often leads to further mobility, insofar as the discovery of Women's Studies is experienced as so enticing that women then seek out ways of obtaining degrees in Women's Studies, which often are not available in their home countries. Rather than going 'back home', these women simply end up going 'elsewhere'. This self-generating effect of mobility can be summarised as follows: 'The desire for mobility mobilizes' (2003, 14) and 'the experience of mobility is likely to make students more mobile' (2003, 27). Those who remain outside the loop of academic travelling are, non-surprisingly, those women who either lack the necessary funds to go abroad and/or are tied up in domestic circumstances (family and/or children) that cannot easily be left behind or taken along.

Where movement takes place, how and where one belongs becomes more complex. The negotiations taking place in the v*ifu* network and the mobile biographies of *ifu* participants can be read through the moments of making belonging, the process of attaching oneself – to others, to a particular set of politics, to a place, an idea, an emotion – that lend a heightened sense of meaning to the very concept of mobility. Or, how one of the *ifu* participants describes this process:

> I have found myself more and more participant in a large loose web of (European) feminists who read each other and think with each other and more and more support each other also emotionally and personally: in short some kind of non-local feminist post-family is forming bit by bit and I am continually amazed to feel so comfortable in it.

Sara Ahmed (2004), proposes that there is an 'instructive' relationship between movement and belonging in so far as 'movement connects bodies to other bodies' and 'attachment takes place [...] through being moved by others' (2004, 11). Like Ahmed, Elspeth Probyn (1996) stresses the affective elements of belonging when she states that 'the desire to belong propels, even as it rearranges, the relations into which it intervenes' (1996, 13).

That is to say, belonging includes affective dimensions, and mobility also happens 'within'. The layers of meaning of mobility are not merely a question of movement but of the making of particular identities, relations to the world and affective attachments with which subjects are implicated in the world. Belonging in the context of feminist networks can take a number of meanings: as negotiations of sameness and difference in the wider feminist community; as issues of dissonance and togetherness; addressing questions of solidarity and its fraught relationship with difference. V*ifu*, for example, is described by one participant as 'part of a global group of women with a shared past and sometimes shared goals, interests, worries'. As such, it is described as 'a place for women, of women, a community'. And despite certain differences, what prevails in such a global platform is 'the common interest in what women are like, are doing, are struggling with and for'. This is what allows for a sense of belonging.

Conclusion

Based on my research findings of a cyberethnographic analysis of the International Women's University 'Technology and Culture' (*ifu*) 2000 and its participants, I explored in this chapter how mobility is experienced and reflected upon, but also created in the construction of togetherness and belonging in feminist online networks. I have argued that such networks as well as the body of feminist theory can be seen as a useful tool in order to understand contemporary mobility, which, especially in the context of the academy, increasingly spans geographical movement as well as the crossing of established theoretical and political boundaries. Understanding the interplay between these two modes, I suggest, can have a fruitful input into better understanding the concept of mobility itself.

Drawing on some of the interactions on the v*ifu* mailing lists, I demonstrated here that virtuality, as an integral part of the event *ifu*, invites us to consider the onsite-event *ifu* and its online counterpart v*ifu* as intimately connected. Moreover, they can be said to constitute each other as places of attachment and detachment based on and generating differing kinds of mobility and belonging. The interaction between participants in the two spheres suggests a more complex relationship between corporeal mobility and virtual mobility than simple binary mirroring. Rather, it points to at least two things, as does the empirical material used throughout this chapter: First, it suggests that equipped with contradicting elements of 'realness' and virtuality, boundedness and flexibility, the v/*ifu* community seems to challenge the notions of belonging and community – imagined (Anderson, 1985) or virtual (Rheingold, 1994) – often associated with cyberspace. V*ifu* is seen as an affective, emotional space based on and perpetuated by an imagined shared experience of having attended the *ifu* event. The shared life experience is perceived to function as the glue that makes list members stick together. Whereas it could and has been argued that emotional online attachment gets established through meeting (like-minded people) in e-spaces (for example Rheingold, 1994) and/or as suggested by Urry through a combination of virtual and face-to-face encounters, the strong emotional attachment to v*ifu* is explained by the shared experience of having participated in *ifu*. This imagination of a common experience complicates not only how we might think about place, interaction, and (online) identity, but the relationship between *ifu* and v*ifu*, as highlighted by the participants' voices throughout, also tells a story of feminism and its relationship to mobility. It tells a story about being-together in different places and interactive modes while taking into account the long established feminist fact that there is no single way of experiencing 'being woman' ('being black' or 'being white', 'being working class' or 'being middle class', 'being European' and so on). In fact, just like Lorraine Code suggests that there is in fact no such thing as a singular 'women's experience' (1988, 190), the negotiations of belonging and being mobile I offered here, always take place on this premise. Thus, being part of this feminist network – both onsite and online – is always constructed around the gaps and frictions that have become so characteristic of feminist theory. Namely, in these negotiations of belonging, 'mobility' is always also discussed and constructed as an intellectual position, as well as movement across theoretical and geographic boundaries, but also as the acknowledgement of the importance of difference and the desire to connect despite and beyond differences.

References

Adams, P.C. and Warf, B. (1997), 'Introduction: Cyberspace and Geographical Space', *Geographical Review* 87, 139–45.

Ahmed, S. (2000), *Strange Encounters. Embodied Others in Post-Coloniality* (London: Routledge).

Ahmed, S. (2004), 'Colletive Feelings Or; The Impressions left by Others', *Theory, Culture & Society* 21, 25–42.

Alarcon, N., Kaplan, C. and Moallem, N. (1999), *Betweem Woman and Nation: Nationalisms, Transnational Feminisms, and theState* (Durham, NC/London: Duke University Press).

Anderson, B. (1983), *Imagined Communities: Reflections on the Origin and Spread of Nationalism* (London: Verso).

Balsamo, A. (1996), *Technologies of the Gendered Body. Reading Cyborg Women* (Durham, NC/London: Duke University Press).

Brah, A. (1996), *Cartographies of Diaspora. Contesting Identities* (London/New York: Routledge).

Braidotti, R. (1990). 'United States of Europe or United Colors of Benetton? Some Feminist Thought on the New Common European Community.' *Differences* 2(3): 109–121.

Braidotti, R. (1994), *Nomadic Subjects. Embodiment and Sexual Difference in Contemporary Feminist Theory* (New York: Columbia University Press).

Braidotti, R. (2002), *Metamorphoses. Towards a Materialist Theory of Becoming* (Cambridge: Polity Press).

Brown, W. (1997), 'The Impossibility of Women's Studies', *Differences* 9, 79–101.

Code, L. (1988), *Experience, Knowledge, and Responsibility. Feminist Perspectives in Philosophy*. M. Griffith and M. Whitford (London: Macmillan) 187–204.

Escobar, A. (1994), 'Welcome to Cyberia. Notes on the Anthropology of Cyberculture', *Current Anthropology* 35, 211–31.

Franklin, S., Lury, C. and Stacey, J. (eds) (2000), *Global Nature, Global Culture. Gender, Theory, Cultural Studies* (London: Sage).

Gajjala, R. (2001), 'Studying feminist e-spaces: introducing transnational/post-colonial concerns', in S.R. Munt (ed.), *Technospaces: inside the new media* (London/New York: Continuum).

Gray, C.H. (ed.) (1995), *The Cyborg Handbook* (New York/London: Routledge).

Griffin, G. and Hanmer, J. (2003), 'Comparative Data Report 8: The Impact of Women's Studies Training on Women's Lifestyles and Everyday Life Practices', http://www.hull.ac.uk/ewsi.

Gunew, S. (2003), 'The Home of Language: A Pedagogy of the Stammer', in S. Ahmed, C. Castaneda, A.-M. Fortier and M. Sheller (eds), *Uprootings/Regroundings: Questions of Home and Migration* (Oxford: Berg).

Gürses, S. (2001), *Putting Technology into Context: Training at virtual ifu* (Internationale Frauenuniversität). Berlin, www.informatik.hu-berlin.de/~guerses/pps/vifuInnsbruck.pdf.

Hannam, K., Sheller, M. and Urry, J. (2006), 'Editorial: Mobilities, Immobilities and Moorings', *Mobilities* 1, 1–22.

Haraway, D. (1991), 'Situated Knowledges: The Science Question in Feminism and the Privilege of Partial Perspective', in D. Haraway (ed.), *Simians, Cyborgs, and Women* (London: Free Association Books).

Hartsock, N. (1998), *The Feminist Standpoint Revisited: And Other Essays* (Boulder, Colorado/Oxford: Westview Press).

Hemmings, C. (2006), 'The Life and Times of Academic Feminism', in M. Evans, K. Davis and J. Lorber (eds), *Handbook of Gender and Women's Studies* (London: Sage).

Kaplan, C. (1994), 'The Politics of Location as Transnational Feminist Critical Practice', in I. Grewal and C. Kaplan (eds), *Scattered Hegemonies. Postmodernity and Transnational Feminist Practices* (Minneapolis/London: University of Minnesota Press).

Kaplan, C. and Grewal, I. (2002), 'Transnational Practices and Interdisciplinary Feminist Scholarship: Refiguring Women's and Gender Studies', in R. Wiegman (ed.), *Women's Studies on its own* (Durham and London: Duke University Press).

Martin, B. (1997), 'Success and Its Failures', *Differences* 9, 102–31.

Massey, D. (1993), 'Power-geometry and a Progressive Sense of Place', in J. Bird, B. Curtis, T. Putnam, G. Robertson and L. Tickner (eds) *Mapping the Futures: Local cultures, global change* (London and New York: Routledge).

Mendoza, B. (2002), 'Transnational Feminisms in Question', *Feminist Theory* 3, 295–314.

Metz-Göckel, S. (ed.) (2002), *Lehren und Lernen an der internationalen Frauenuniversität. Ergebnisse der wissenschaftlichen Begleituntersuchung* (Opladen: Leske and Budrich).

Mihel, J.S. (2002), 'Revising Sisterhood: From Unity to Affinity. Identification Processes in the Context of Globalized Communication - The Case of the International Women's University', in C. Floyd, G. Kelkar, S. Klein-Franke, C. Kramarae and C. Limpangog (eds), *Feminist Challenges in the Information Age: Information as a Social Resource* (Opladen: Leske and Budrich).

Mohanty, C.T. (2003), *Feminism Without Borders: Decolonizing Theory, Practicing Solidarity* (Durham/London: Duke University Press).

Morley, L. and V. Walsh (1995), *Feminist Academics: Creative Agents for Change* (London, Taylor & Francis).

Nakamura, L. (2002), *Cybertypes: Race, Ethnicity and Identity on the Internet* (New York and London: Routledge).

Neusel, A. (ed.) (2000). *Die eigene Hochschule. Die Internationale Frauenuniversität 'Technik und Kultur'.* Schriftenreihe der Internationalen Frauenuniversität 'Technik und Kultur', Band 1. Opladen, Leske + Budrich.

Plant, S. (1997), *Zeros + Ones: Digital Women + The New Technoculture* (London: Fourth Estate).

Probyn, E. (1996), *Outside Belongings* (New York/London: Routledge).

Rheingold, H. (1993), *Virtual Community: Homesteading on the Electronic Frontier* (New York: Addison-Wesley).

Rheingold, H. (1994), *The Virtual Community: Finding Connection in a Computerized World* (London: Secker & Warburg).

Rich, A. (1984), 'Notes Toward a Politics of Location', in A. Rich (ed.) *Blood, Bread, and Poetry. Selected Prose 1979 - 1985* (London, Virago).

Schelhowe, H. (2001), 'Offene Technologie - offene Kulturen. Zur Genderfrage im Projekt Virtuelle Internationale Frauenuniversität (vifu).' FifF Ko 3: 14–21.

Schor, N. and E. Weed (1990), 'Notes from the Beehive: Feminism and the Institution.' *Differences: A Journal of Feminist Cultural Studies* 2.

Scott, J.W. (1997), 'Women's Studies on the Edge. Introduction', *Differences* 9, i–v.

Skeggs, B. (1995), 'Women's Studies in Britain in the 1990s: Entitlement Cultures and Institutional Constraints', *Women's Studies International Forum* 18, 475–85.

Spivak, G.C. (1996), 'Diasporas Old and New: Women in the Transnational World', *Textual Practice* 10, 245–69.

Star, S.L. (1996), 'From Hestia to Home Page: Feminism and the Concept of Home in Cyberspace', in N. Lykke and R. Braidotti (eds) *Between Monsters, Goddesses and Cyborgs* (London: Zed Books).

Stone, S. (1991), 'The *Empire* Strikes Back: A Posttranssexual Manifesto', in J. Epstein and K. Straub (eds), *Body Guards. The Cultural Politics of Gender Ambiguity* (New York and London: Routledge).

Turkle, S. (1996), *Life on the Screen: Identity in the Age of the Internet* (London: Weidenfels and Nicolson).

Urry, J. (2002), 'Mobility and Proximity', *Sociology* 36, 255–74.

Urry, J. (2003), 'Social Networks, Travel and Talk', *British Journal of Sociology* 54, 155–75.

Wiegman, R. (1999/2000), 'Feminism, Institutionalism, and the Idiom of Failure', *Differences* 11, 107–36.

Wiegman, R. (ed.) (2002), *Women's Studies on its Own: A Next Wave Reader in Institutional Change* (Durham/London: Duke University Press).

Yuval-Davis, Nira (1997), *Gender and Nation* (London/Thousand Oaks: Sage).

Chapter 6

Gendering Mobility: Insights into the Construction of Spatial Concepts

Nadine Cattan

Introduction

Understanding how mobility alters relationships between societies and space and calls for a re-assessment of spatial theories and concepts is in itself a fairly complex field of research. To add the exploration of gender to these interconnections between mobility and geography is a real challenge.

Over the last 20 years a large body of work – often extending well beyond the field of geography – has underscored the importance of these issues. Some of this work has set out to assert the importance of these research themes in academia (Castells, 1996; Veltz, 1996; Taylor, 1999; Law, 1999; Sassen, 1999; Longhurst, 2002; Boyle, 2002), while others have engaged in more in-depth studies, attempting to construct theoretical and conceptual arguments concerning gendered mobility patterns that could fundamentally change our modes of thought (Hanson and Pratt, 1988; Pratt, 1997; Bondi and Rose, 2003; Tyner, 2003; Silvey, 2006; Cresswell, 2006; Sheller and Urry, 2006).

On the basis of this wealth of literature,[1] and specifically two studies of student mobility in Europe and on home/workplace mobility in the Paris metropolitan area, this chapter is intended to show why gendering mobility provides new theoretical advances in the construction of spatial concepts and spatial representations. With this as the context, I will identify the reasons behind the difficult relationship between geography and mobility. Then, using selected literature, I will focus on the opportunities and conceptual advances that are generated by the interaction of gender, mobility and geography. Particular attention will be paid to the way in which a gendered approach to mobility enables the reinterpretation and reassessment of current prevalent concepts.

Rather than offering a normative approach, the discussion in this chapter highlights the importance of the gender dimension in the construction of a *critical* understanding of our spatial concepts, and our interpretation of the spatial behaviour of populations.

---

1 This analysis refers solely to publications in English or French, where empirical studies have been conducted in the West.

Mobility and geography: a difficult relationship where gender requires a hearing

Today it is commonly agreed that space entails networks. However, numerous authors have pointed out that consideration of mobility as an issue has not yet been fully integrated into theories of space and its dynamics, and have argued for a reading of spatial organisation via exchanges, flows and mobility patterns.[2] To cite only a few among the most radical, Sheller and Urry (2006) underscore the fact that much social science research has been 'a-mobile', and Cresswell (1996) notes that the study of mobility has not been accorded the same attention as the study of place, space, territory, boundary and landscape.

The reasons given by many scientists to explain why it is difficult to take the realities of mobility into account consistently relate to material or technical contingencies, ranging from lack of access to relational data to the methodological complexity of using such data. It is true that these limitations are considerable. Yet the debate is incomplete, while at the same time it seems that any attempt to give meaning to space and to populations in terms of linkages and interdependencies, rather than in terms of zones and distributions, meets resistance in various forms: symbolic, ideological, and institutional.

The weight of symbolic and ideological preconceptions

The issue addressed here is the weight of dominant interpretations of territorial, and in particular urban, realities. I do not intend to give a detailed description of mobility patterns in contemporary societies, or an assessment of their magnitude and evolution, nor do I intend to adopt a dualistic standpoint as to whether or not these patterns should be controlled. Instead, I will attempt to analyse the symbolic, ideological and cultural load weighing on the interpretation of 'mobile territorialities', and hence on our representations of these territorialities.

There are four major sources of inertia, which, more than others, impede the renewal of social scientific thought regarding mobile territoriality and populations in movement.[3] *Post-hoc* interpretation of these forms of resistance shows that they all have a direct or indirect link with gender, as I will explain below.

The first source of inertia is cultural, and relates to an as-yet unresolved conflict in perceptions of network and territory. Mobility, and the networks that underpin mobility, are seen as disconnecting populations from their territories, and in particular from their everyday living environments. Mobility and networking indeed contribute to deconstruct a perception of territories, and particularly of European urban spaces, structured by distance according to a regular and predictable relationship that reflects a decrease of densities along a gradient from the centre towards the peripheries. With

2 Theoretical and conceptual argumentation is provided in particular by the following authors: Castells, 1996; Wittel, 2001; Dupuy, 1991; Veltz, 1996; Claval, 1981; Raffestin, 1980; Taylor, 1999; Cattan, 2004; Sheller and Urry, 2006; Cresswell, 2002; Clifford, 1992.

3 These factors have also been discussed in a joint publication by N. Cattan and S. Berroir on urban sprawl published in 2006.

mobility, a new kind of space is emerging, a space made of movement, a space which itself becomes a network where the gap is unceasingly recreated due to a dynamic where lines and tunnel effects are observed. Consequently, mobility and networks have the capacity of deterritorialisation, and this reality is frightening, because it creates fragmented identities and spaces. In this context, mobility and networks are seen as a threat to the spatial and social cohesion of territories. This resurrects certain age-old fears, as if in the face of a threatened catastrophe. Indeed, the geographic scientific corpus, which today reflects a modern structuralist interpretation of living, moving and producing, built around notions of centrality, hierarchy, proximity, centre/periphery, and local/global, is at risk of being destroyed, or at least strongly challenged. These concepts are not only fundamental academic references in social sciences. Some, such as centrality, hierarchy and the centre/periphery model, constitute symbolic elements associated with a whole system of cultural and metaphysical values.

This first source of inertia is compounded by a second, which is related to our own intellectual limitations, whether in terms of conceptual references or in terms of imaginative ability. It appears that apprehending a mobile reality is a challenge to our imaginations. In as much as it is firmly rooted in a binary rationality of modes of thought, we seem to have difficulty reflecting on place and circulation, immobile and mobile, compact and fluid not merely as two complementary elements, but as pairs that form a single entity (Chalas, 2000). Pierre Veltz (1996) underscores the fact that our representations extrapolate traditional models of territory, and miss the real novelty, which is that a territory of networks is replaced by a networked territory.

The third source of inertia is rooted in a conception of societies in which order and power occupy a central position. Unlike the impalpable and unpredictable realities of a territory in continual movement, a well-delineated space gives the impression that it is managed and structured. It offers the gratifying illusion of controlling the populations living within it. Mobile territorialities, although markedly channelled by the material existence of massive transport infrastructures, are nonetheless fluid and ever-changing. This alarms those in positions of responsibility, whether in the field of development, spatial planning or territorial management.

Finally, it seems to me that there is a fourth inertial factor weighing even more heavily than the other three on our willingness to envisage mobility in its full scope. To envisage mobility as a factor that actually produces social and spatial identities is to topple the ultimate symbol of our mainly urban contemporary civilisation, that of sedentarism. Few people today are prepared to take this step. After all, the history of civilisations has been constructed around the remains of the first sedentary settlements dating from the Neolithic, and rendered possible by agriculture. As Malkki (1992; 27) points out, 'the widely held commonsense assumptions linking people to place, nation to territory, are not simply territorializing but deeply metaphysical'. Thinking about nations and national identities may often take the form of roots: culture is conceived of as something inextricably linked to the soil, and people think of themselves as being rooted in place. Being out of place implies being unable to fit in society (Cresswell, 1996; Sibley, 1999). Thus, the metaphysics of sedentarism enables the consolidation of a nationally bounded geography that consolidates the world order (Smith, 1986). To reconcile the history of our modern civilisations with components of nomadic societies amounts to questioning the very foundations of our modern history. It is a question of rehabilitating the temporary and the ephemeral (Augé, 1995; Cresswell, 2006).

In a general manner, it is as if mobility threatens to strike down the theoretical, symbolic and identity values of our modern societies. However, to trade the myth of the sedentary lifestyle, border and frontier, territory and proximity, for the myth of mobility, flows and networks would be a mistake. It is important to realise the strength of symbols, and to avoid replacing one strong symbolic reference with another, even if it is hybrid and constantly changing.

Mobility: a hybrid concept that is gender-related

The concept of mobility has been defined in a number of ways. In its most frequent formulation, mobility is displacement, in real space or in virtual space, of people and objects. In this definition, mobility is not restricted to physical displacement – it also integrates the mobility potential arising from intentionality on the part of individuals, from their strategies, and from the arbitrations that underpin this (Kaufmann, 2002; Urry, 2000; Kesselring, 2006). Displacement also changes the person (or object) involved in it. The idea that spaces, as well as people, change their nature to themselves become mobile entities, runs through a large body of study.

The present rapid evolution of mobility behaviours leads us to reassess dominant representations and concepts regarding the relationships between populations and territory. Territorial constructs are no longer experienced and defined in terms of expanse and limits formed by links of proximity within a spatial continuum. They become reticulate, as a result of connective relationships that develop between distant places. They become topological. In this context, far and near are no longer contrasted in a binary classification of space, and an expansive terminology has emerged over the last fifteen years to refer to the plurality of territorial identities. It is, for instance, common to refer to 'nomadic' societies to ground this change in perspective. The image of the archipelago also gives a good picture of these dynamics, while notions such as 'circulatory territories' and 'circulating populations' (Tarrius, 1994) explore even further the effects that mobility patterns may have on societies.

These mobility patterns are often associated with those who are better off in Western societies, and mainly with white middle-class men in rich countries. Yet a large number of feminist geographers, sociologists and economists show considerable interest in the mobility theme (Hanson and Johnston, 1985; Hanson and Pratt, 1988; Villeneuve and Rose, 1988; Clifford, 1992; Cresswell, 2002; Tyner, 2003; Silvey, 2004; Cattan, 2007). In the approach adopted by these authors, mobility is often treated as a metaphor reflecting new relationships with here and there, with others, and *in fine* with otherness. Being mobile is not just about geographical space, but also, and probably above all, about social space. By placing the social dimensions of mobility at the centre of debate, these studies have been less concerned with describing the spatial forms of mobility networks than with determining the socio-spatial meanings and consequences of this mobility. This explains why gender studies take such an interest in mobility. Indeed, as mobility becomes a metaphor suggesting a quite different relationship with distance, and as space and consequently distance relate to society and to power, then mobility is seen as a way to rethink and to deconstruct the framework through which social control is expressed.

In relation to the four types of inertia that carry weight in shaping perceptions today, mobility undermines the view of territories as being ordered according to a regular and predictable pattern. These questionings, directly or indirectly, imply a re-assessment of the prevalent socio-economic and geographic theory where the notion of distance is fundamental. Several studies explain why Euclidian geometry provides the basic representation of objective space that corresponds to a spatially ordered landscape (Harvey, 1989) and why distance is viewed as a powerful factor in the structuring of our lifestyles, our modes of living, and our modes of displacement and production. By way of their conception of time-space, these classic approaches interpret socio-territorial organisation patterns through the prism of proximity and spatial continuity. Yet such interpretations lead to a static apprehension of territories and societies in terms of dual categories such as local/global, here/there, periphery/centre, inner/outer, or private/public. In these dual categories, the first term is often associated with the feminine sphere, and the second with the masculine sphere. In that, the notion of distance and Euclidian space, consciously or otherwise, carries a fixed meaning. Consequently, it is as if removing distance as a key to interpreting our movements and our territories amounts to undermining the fundamental dominant concepts upon which our social scientific knowledge, and beyond it our social models, are constructed. If the concept of 'time-space compression' questions the very notion of distance and in particular its Euclidian dimension, it deconstructs in a very moderate way the classic conceptions of space. Indeed, time-space compression continues to convey an image of space considered as a continuum, and consequently it sets its continuous dimension in opposition to its discontinuous dimension. The second part of this chapter, by re-exploring the relevance of the notion of distance, shows that mobility could in fact contribute to a more 'hybrid' approach of spatial concepts, one that is reinforced when mobility is analysed according to gender.

Gender in mobility: a theoretical concept beyond the dominant representations of space and its organisation

The interest in gender and mobility has two important facets. The first is the issue of the impact of mobility on gender, which leads to the questioning of the explicative factors of spatial behaviours as differentiated according to gender. The second aspect questions the effects of gender on mobility, which leads to re-interpretations of spatial categories and concepts.

Starting with two major fields in the literature on gendered mobility, that is home-to-workplace commuting and international migrations, this work aims to show how gendering mobility is not only at the forefront of new critical interpretations of spatial categories, but also allows us to generate concrete theoretical advances on the subject of contemporary change in territorial functioning and the representations we have of it. The picture I am going to set out is far from exhaustive. It gives a partial view of certain major aspects in connection with these opportunities and advances found in the literature.[4] This chapter will focus on: a) the spatial reinterpretations of gendered mobility patterns, b) the re-assessment of certain dominant assumptions concerning spatial processes, and c) new perspectives on certain important spatial objects.

Modifying the interpretation of spatial behavioural differences

A large part of the literature on gendered spatial issues analyses different travel behaviours of men and women, with the purpose of explaining them. The most frequent theme is the distance of the daily commute to work in North American and European metropolitan areas. On average, commuting distances are reported to be shorter for women than for men (Madden, 1981; Fagnani, 1983; Hanson and Johnston, 1985; Hanson and Pratt, 1988; Villeneuve and Rose, 1988; Thomas and Villeneuve, 1998; Lemelin and Gatignol, 1999; Cattan, 2007).

In this respect, this first body of work requires attention, since it questions a classic hypothesis according to which the shorter home-to-work commuting distances observed for women are mainly explained by their broader family and domestic responsibilities, in spite of the fact that other explanatory factors have been evidenced (for an overview, see Blumen, 1994 and Hanson and Johnston, 1985).

Hanson and Johnston (1985) showed in Baltimore, Maryland, that family responsibilities (assessed via the presence of children in the household and according to age) did not contribute to explaining the differences in daily commuting behaviours between women and men. By identifying twelve different household profiles, they also showed that the presence in the household of a pre-school child does not significantly alter female mobility patterns. Thomas and Villeneuve's (1998) work on the city of Quebec provide similar conclusions. It showed firstly that women who have the youngest children travel the greatest distances to reach their workplace, and secondly that single mothers travel greater distances than single women living alone. By showing that in the Paris metropolitan region, women with children commute further than women without children, and that women with young children on average travel greater distances than women with older children, my own work confirms the preceding conclusions (Cattan, 2007).

Consequently, these results have led to the search for other explanations for gender mobility patterns in the public sphere, rather than solely personal and private determining factors. Several works have shown that mainly female employment zones are very different from mainly male employment zones, and also that these zones concentrate in very specific spatial sectors in the urban regions studied. According to Massey's now famous phrase, these observations underline the fact that 'space matters'. This interpretation of the determining factors in migratory behaviours supports theories on the spatial segregation of employment opportunities according to gender, very well summed up by two highly evocative notions, that of the 'captive rider' (Rutherford and Wekerle 1988, 116) and that of 'spatial entrapment' (Hanson and Pratt, 1994, 196), both of which highlight the lack of choice and opportunity

4 This literature is mainly geographical, and is in English or French as noted in the introduction.

in spatial terms for working women. The relative spatial containment of women in metropolitan areas vis-à-vis the labour market access is also theorised through the spatial mismatch hypothesis that reinforces sex segregation. A thorough analysis of the extent of the local labour market in several metropolitan areas (Wyly, 1998, McLafferty and Preston, 1992, Hanson and Pratt, 1994) shows how gender and ethnic dimension are interdependent in the complex patterns of spatial mismatch. The findings suggest indeed that gender, ethnicity and class are mutually constituted additive social divisions (McDowell, 1993; McLafferty and Preston, 1992, 429). Thus in all these studies the explanations of women's spatial containment rest upon a structural perspective and contrast with the dual-role interpretation of women's shorter home-to-work commuting distance (Rutherford and Wekerle, 1988).

There is a second body of work that I would like to focus on, because it contributes to questioning the relevance of a second factor that is often used to explain the shorter commuting distances for women, which is socio-professional position, evidenced via income, educational level or socio-professional category. Indeed, if a large number of studies show that the considerable proportion of women in low-income employment contributes to explaining their shorter average commuting distances, some studies show that for any given income women still commute over shorter distances than men (Villeneuve and Rose, 1988; Hanson and Johnston, 1985; Cattan, 2007). Indeed, in Montreal, Baltimore or Paris, the distance commuted from home to work by women in a given socio-professional category is closer to that of women in a different socio-professional category than to that of men in the same category. The authors show, in addition, that these gender differences in mobility behaviours are greater for the 'higher' socio-professional categories. These results contrast with classic hypotheses, which continue to assert that socio-economic and professional affinities are generally stronger than those of gender (Lemelin and Gatignol, 1999). Further, they question certain assumptions of urban ecology, and suggest a reinterpretation of the processes underpinning daily commuting to work.

Renewing the idea of spatial 'processes'

In the second part of this exploration of gender and mobility, the issue is to understand how comparing the mobility patterns of men and women contributes to a reinterpretation of spatial theory, in particular theory relating to spatial integration and globalisation.

Most of the work on spatial integration and globalisation focuses on what is known as structuring flows, such as financial flows, commercial exchanges, freight or commodity flows. Thus these studies focus on the functional integration of economic activities across the globe. In such analyses, the conceptualisation of spatial integration and globalisation by way of networks of cities has often been restricted to two models derived from classic representations of urban systems at the national level, that is, the centre-periphery model and the hierarchical network model. These representations consider distance as the main principle of spatial organisation. They express the idea of a gradient of a progressive decrease of concentration, capacity, power and richness from the centre towards the peripheries. Consequently, the position of a territory in relation to this centre has many consequences. As a result, space and its organisation at World or European levels, for example, is seen in a dichotomous manner: either as a dominant centre to which dependant and isolated peripheral areas are more or less well attached, or else as a set of major poles with secondary satellite centres that suffer from a lack of visibility (Cattan and Saint-Julien, 1998). Over the last ten years or so, a large number of studies specifically point out that other types of flows deserve consideration, and that they are often neglected in the literature. Examples of these are activist networks concerned about environmental and health issues, or trans-national networks of ethnic or women's groups (Nagar et al., 2002; Sassen, 1999; Marchand and Ruyan, 2000).

In an approach aiming to go beyond static representations of the integration of European territorial systems that integrate only wealth-producing processes I conducted a study of male and female student migrations in Europe.[5] Student exchanges are a particular migratory phenomenon, because the migrations occur within a time scale that is relatively shorter than that of other migratory flows, and because the decision to migrate results from personal factors.

In the year 2000 about 109,000 students moved to another country for reasons of study, and 61 percent of these migrants were women. The research I conducted revealed that female students, everywhere in Europe, are, all else being equal,[6] significantly more mobile than male students, and that medium-sized cities were the preferred destinations chosen by female students (Cattan, 2004). In the absence of socio-economic surveys, it is only possible to hypothesise explanations for the different motivations behind these migratory patterns. On the negative side, it is possible to suggest reluctance or fear in relation to a large city. Conversely, medium-sized cities may exert positive attraction, since they are often viewed as being more human in dimension, and thus may meet expectations of quality of life, living environment and lifestyle that women students may integrate into their decisions to a greater degree than men. In these spatial choices by students, it is possible to see the emergence of a new model of female migrations (Balding et al., 1996): female students weary of the prevalence of the social model based on economic competition would seek alternative ways of life, where cultural aspects are better integrated into daily life.

Whatever the reasons for their choice, the scope for alternative forms of territorial development can be seen in this gendered selection of urban destinations. This hypothesis is reinforced by the migratory itineraries that are preferentially engaged in by female students. Indeed, the most predominantly female inter-urban migration routes show a network characterised by very weak polarisation, in which the favoured associations correspond neither to an urban logic, nor to a specific spatial logic (Cattan, 2004a). In Europe, female student migrations, to a greater extent than male student migrations, contribute to ongoing spatial re-organisations, and to the construction of alternatives to metropolitanisation. These migrations could reflect more flexible spatial configurations in which more polycentric, more reticulated and possibly ultimately fairer territorial development patterns can form. Indeed, because

5 Mobility within the ERASMUS programme.

6 In particular regarding the numbers of students differentiated according to gender in each of the European countries studied.

cities and networks are in the neo-classic spatial model positioned within a logic of economic and political competition in which only processes of wealth generation are taken into account, the conceptualisation of territorial dynamics is always viewed in a hierarchical, pyramidal manner. These conceptualisations of spatial processes place great emphasis on the size of places and to their relative geographical position (that is their distance one to each other). In these views, the large metropolises have a particular statute and become the only ones that matter. By showing that the female student inter-urban migration network is polycentric and balanced, that is that no city or pair of cities dominates the network of exchanges, this study highlights the fact that the migratory behaviour of female students goes further in questioning these neo-classic hypotheses of spatial mobility than do those of male students.

In the studies on globalisation that take mobility and gender into account, similar concerns are to be found. Numerous studies have indeed pointed out that the images of globalisation that are traditionally fostered are based, to a large part, on mass criteria. Because they restrict themselves to the analysis of the strongest flows and the largest poles, the world spatial vision generated by these studies is often fixed and inert, always offering the same typologies and the same classifications of international cities and flows (Kelly, 1999; Sparke, 2001; Cattan, 2004a). Places in the South are viewed, if indeed they are given any consideration at all, as 'mere recipients of globalisation', rather than being able to act upon it. Yet today the increasingly complex interweaving of situations of centrality with peripheral situations renders most of the dominant models less operational, if not obsolete. Like the countries of the South, women are also marginalised in much of the literature on globalisation.

Additionally, by focusing on a macro-scale reading of the economic and social spheres, most of the studies on globalisation place the concept of the nation-state at the centre of debate. This global/national focus reinforces notions of space as being discrete and bounded, and works against taking account of trans-national experiences, or processes emanating from finer scales such as households or local communities (Nagar et al., 2002; Escobar, 2001; Sassen, 1999; Silvey, 2004). Highlighting the influence of women, or of cities in the South, or any of the spatial and social levels at which globalising processes occur, makes it possible to break away from the constant reiteration of a single dominant view of the world, essentially masculine and West (self) centred, which has been so widespread that it has led us to see only the visible part of the iceberg.

It is, above all, the work on trans-national migrations that has brought new life to the debate. This renewal takes the shape, for instance, of exploration of the migrations of working women as domestic workers. In non-feminist research on migrations, this type of work is regarded as private and often neglected, or considered to be less important than salaried work in more formal economic sectors (Pratt and Yeoh, 2003). More generally, low-wage jobs, mainly held by women, do not correspond to our representations of globalisation. Yet they are a genuine part of the process, and Sassen (1999) underscores the fact that economic globalisation needs to be understood not only in its macro-level dynamics, but also in its multiple localisations. These multi-level approaches integrate the scales of body, household, community, and the trans-national dimension, and emphasise above all the way in which the different scales are formed and conceptualised in relation to the social, political and cultural arenas (Nagar et al., 2002; Cox, 1997; Kelly, 1999; Katz, 2001).

In revising gender-neutral approaches to globalisation, feminist migration research has contributed to explaining the role of difference in socially constructing spatial scales, as well as in the various meanings of the scales through which the dynamics of the causes and consequences of migration are played out (Silvey, 2004).

Challenging the definition of spatial 'objects'

The third critical issue is the redefinition of several spatial 'objects' in studies on gender and mobility. Over the last fifteen years or so, feminist geographers have been re-examining the notions of space and place. The themes in which they are engaged are among those that are the most likely to open up new lines of study. Moving beyond strongly economist approaches to the understanding of space, where place is bogged down by representations of borders and limits, involving notions of inside and outside, the analysis of migrations in relation to gender has led to the definition of principles for the conceptualisation of space in terms of space-time, formed from social interrelations on all scales. From this perspective, 'the identities of place are unfixed, contested and multiple. They are not constructed in relation to the other which lies beyond, but precisely through the specificity of the mix of links and interconnections to that *beyond*' (Massey, 1994, 5). Today, in a more general manner, the explorations conducted by certain researchers on the city have integrated these conceptual advances, advocating an urban theory constructed around the notion of the 'emerging city', that is a city that is viewed as a non-fixed form in constant evolution, a city that is at once mobile and territorial, simple and multiform, polycentric and empty (Dubois-Taine and Chalas, 1997), a city with plural identities in which ideal models have no place (Mongin, 1995).

While attempts to reconcile spatial and social components that are classically viewed as being antagonistic are not the prerogative of gender studies, it is nevertheless clear that these studies have questioned the classic theories of space, in particular by their very explicit reference to the notions of identity and power. In their work on women in the Third World, Mohanty et al. (1991) apprehend place and community through a relational approach, based on the logics of opposition towards the established power. This point of view is interesting in the sense that it leads us away from a spatially fixed, bounded notion of space, place or community that is based, for example, on mutual co-residence. Instead it suggests a political basis for defining our objects of study (McDowell, 1993).

To remove place from its static assumptions is not merely academic debate. To position place and space in a dynamic relationship with societies and gender amounts to attempting to see social and spatial identities as being plural, and not reducible to a single model. Indeed, our definitions of spatial objects need to integrate a hybrid dimension.

Similarly, a certain number of studies on gendered mobility propose to revise the concept of scale. Because they focus on the effects of social constructions of spatial objects, these studies question the ontological definitions of spatial scales (Cox, 1997; Marston, 2000). Basing their work mainly on international migrations of women, gender studies address questions about the political or gendered processes underlying the construction of scale, unlike neoclassic migration research 'which considers scale as a spatial category within which migration plays out, and not as a focus of enquiry' (Silvey, 2004, 4).

In so doing, they show how overcoming segmented understandings of scales requires a more integrative perspective, progressing from body to household to region, to nation and to the trans-national dimension (Massey et al., 1993; Silvey, 2004). The questioning of the national scale and the meaning given to it in classic approaches to international migrations is the most emblematic aspect. Indeed, as seen in the previous section, studies on women migrating as domestic workers, for example, show the partiality of the definition of the national scale, which derives from dominant social and political constructions, focusing on certain dimensions and groups, and excluding others considered marginal or secondary (Yeoh and Huang, 1999; Katz, 2001). In an attempt to understand the mechanisms underpinning gendered migrations from countries like the Philippines, Sri Lanka or India, a large number of studies underscore the political nature of mobility and the social dynamics connected to it, looking at both ends of the process – countries receiving immigration and those generating emigration (Pratt, 1997; Tyner, 2003; Sassen, 1999; Katz, 2001; Silvey, 2006). This means that unlike classic approaches, which explore 'the way in which gender hierarchies are constituted as a part of the way women and men learn to identify with a nation state', gender studies explain 'how gender and nation are mutually constituted within the trans-national social field that links places' (Boyle, 2002, 535).

Gender studies also show that two scales that are often neglected in classic theory, that of the household, and more recently that of the body, contribute to the way in which the different territorial levels, and in particular the national level, are constructed. By highlighting the social cost of mobility, and by developing the notion of embodiment to refer to gendered processes that at present govern access to public spaces and transport, for example, these studies show how subjective our definitions of scale can be, as well as the modes of regulation that go with them (Cresswell, 1996; Bondi and Rose, 2003; Silvey, 2006).

By showing how the different scales are systematically defined in a manner that disconnects them from gendered mobility patterns, proponents of gender study deny the conceptual relevance of spatial objects as being culturally masculine, in particular those objects relating to scale and space. Massey (1994), however, does caution against replacing these 'masculine' viewpoints with 'feminine' viewpoints. She calls for the development of research on the links between the current culturally specific construction of gender and certain aspects of our conceptualisation of spatial 'objects'. It is in this direction that gender studies needs to move to meet the challenge of a critical reading of geographical knowledge.

Conclusion

Mobility has the potential to renew representations of space and its organisation. It is mainly because it questions the relevance of economic, social and spatial theory, where distance and scale are viewed as strong structuring factors, that mobility makes it possible to go beyond the dualistic categories of our modes of thought and know-how. Indeed, in the majority of the current dominant theories, space is reduced to a geometrical dimension organised by simple elements that can be united and easy to handle (Raffestin, 1980).

This chapter shows how by conveying a conception of space in movement, that is fluid, unbounded and unfixed, mobility deconstructs the classical perception of distance and scale, diluting the very notion of space. The main consequence is that today different forms of territorialities are experienced and can be captured through the image of the archipelago. This image conveys more than other images do a renewed approach of spatial and social identities that question the very basis of the current established order in terms of social control and of power. By and large, this chapter underscores that mobility topples certain major symbols and values of our contemporary Western societies, which are to a great degree related to the urban nature of modern civilisations.

These theoretical and conceptual advances that are being derived today from an examination of mobility and gender give full meaning to the idea that 'writing difference' is concerned with movement rather than position (Doel, 1994). The fluidity of mobile realities and identities, often associated with the idea of post-structuralist 'hybridity', does indeed make it possible to 'write difference', and to go beyond the dominant static conceptualisations of socio-spatial theory.

By questioning certain assumptions of urban ecology regarding the factors that explain gendered behaviours, by showing the narrowness of prevalent representations of spatial processes, and by reassessing the definitions of some spatial objects that are considered to be major references in geography, this study also shows that writing about gender differences in mobility can advance the dialogue between materialist political economy and critical social theory (Silvey, 2004).

This dialogue is a major challenge for social sciences, and the intellectual approach still needs definition. In a satirical attempt to overstep the narrow and controversial frames in which contemporary thought is still confined, Donna Haraway (1991) draws inspiration from 'Actor-network theory' to give us a cyborg creature, half-human, half-machine, at once social reality and pure fiction, to express the process of hybridisation operating everywhere today. Numerous novel theoretical grids are at present emerging to demonstrate the need to think outside, over, and away from the dualistic and antagonistic categories of modern thought. Our researches urgently need to make use of them.

References

Augé, M. (1995), *Non-places: An Introduction to an Anthropology of Supermodernity* (London: Verso), 18.

Balding, V., Euler, C., Hanmer, J., Wigglesworth, D. (1996), 'La mobilité des femmes: migration, citoyenneté et processus d'intégration des femmes en Europe', *Rapport Commission Européenne.*

Blumen, O. (1994), 'Gender differences in the journey to work', *Urban geography* 15 (3), 223–45.

Bondi, L. and Rose, D. (2003), 'Constructing gender, constructing the urban: a review of Anglo-American feminist urban geography', *Gender, Place and Culture* 10, 229–45.

Boyle, P. (2002), 'Population geography: transnational women on the move', *Progress in Human Geography*, 26, 4, 531-543.

Castells, M. (1996), *La Société en Réseaux. L'ère de l'information* (Paris: Fayard).

Cattan, N. (2004a), 'Genre et mobilité étudiante en Europe', *Espace, Populations, Sociétés* 1, 15–27.

Cattan, N. (2004b), 'Le monde au prisme des réseaux aériens', *Flux* 58, 32–43.

Cattan N. (2007), 'Genre, Mobilité quotidienne et emploi en Ile-de-France', Gavray C. (ed.), *Femmes et mobilités* (Bruxelles: Cortex).

Cattan, N. and Berroir, S. (2006), 'Les représentations de l'étalement urbain: un essai d'interprétation', in A. Berque, Ph. Bonin, C. Ghorra-Gobin (eds), *La ville insoutenable* (Paris: Belin, Mappemonde).

Cattan, N. and Saint-Julien, Th. (1998), 'Modèle d'intégration spatiale en Europe', *L'espace géographique* 1, 1–10.

Chalas, Y. (2000), *L'invention de la ville* (Paris: Anthropos, Villes).

Claval, P. (1981), *La logique des villes* (Paris: Litec).

Clifford, J., (1992), 'Traveling cultures', in L. Grossberg, C. Nelson and P. Treicher (eds), *Cultural studies* (New York: Routledge).

Coutras, J. and Fagnani, J. (1978) 'Femmes et transport en milieu urbain', *International Journal of Urban and regional Research* 2, 433–9.

Cox, K. (1997), *Spaces of globalization* (New York: Guilford Press).

Cresswell, T. (1996), *In Place/out of place: Geography, Ideology and Transgression* (Minneapolis: University of Minnesota Press).

Cresswell, T. (2002), 'Introduction: theorizing place', in G. Verstraet and T. Cresswell (eds), *Mobilizing Placing, Placing mobility* (Amsterdam: Rodopi).

Cresswell, T. (2006), *On the move: mobility in the modern Western world* (New York: Routledge).

Doel, M.A. (1994), 'Writing difference', *Environment and Planning A* 26 (7), 1015–20.

Dubois-Taine, G. and Chalas, Y. (1997), *La ville émergente* (La Tour d'Aigues: Editions de l'Aube).

Dupuy, G. (1991), *L'urbanisme des réseaux* (Paris: Colin, U-Géographie).

Escobar, A. (2001), 'Culture sits in places: reflections on globalism and subaltern strategies of localization', *Political Geography* 20.

Fagnani, J. (1983), 'Women's commuting patterns in the Paris Region', *Tijdschrift voor Econ. En Soc. Geografie* 74 (1), 12–24.

Hanson, S. and Johnston, I. (1985), 'Gender differences in work-trip length: explanations and implications', *Urban Geography* 6 (3), 193–219.

Hanson, S. and Pratt, G. (1988), 'Reconceptualizing the links between home and work in urban geography', *Economic Geography* 64 (4), 299–321.

Hanson, S. and Pratt, G. (1994), 'Spatial dimension of the gender division of labor in local labor market', *Urban Geography* 15 (3), 180–202.

Haraway, D. (1991), *Simians, Cyborgs and Women: The Reinvention of Nature* (London: Free Association Book).

Harvey, D. (1990), *The Condition of Postmodernity* (Oxford: Blackwell Publishing).

Katz, C. (2001), 'On the grounds of globalization: A topography for feminist political engagement', *Signs: Journal of Women in Culture and Society* 26.

Kaufmann, V. (2002), *Re-thinking mobility, Contemporary Sociology* (Aldershot: Ashgate).

Kelly, P. (1999), 'The geography and politics of globalization', *Progress in Human Geography* 23, 379–400.

Kesselring, S. (2006), 'Pioneering mobilities: new patterns of movement and motility in a mobile world', *Environment and Planning A* 38, 269–79.

Law, R. (1999), 'Beyond 'women and transport': towards new geographies of gender and daily mobility', *Progress in Human Geography* 23 (4), 567–88.

Lemelin, A. and Gatignol, A. (1999), 'La structure des déplacements pendulaires des femmes et des hommes dans la région de Montréal', *Cahiers de Géographie du Québec* 43 (119), 187–210.

Longhurst, R. (2002), 'Geography and gender: a 'critical' time?', *Progress in Human Geography* 26 (4), 544–52.

Madden, J.F. (1981), 'Why women work closer to home', *Urban Studies* 18, 181–94.

Malkki, L. (1992), 'National Geographic: The rooting of Peoples and Territorialization of National Identity Among Scholars and Refugees', *Cultural Anthropology* 7 (1), 24–44.

Marchand, M. and Runyan, A. (eds) (2000), *Gender and Global Restructuring: Sightings, Sites, and Resistances* (New York: Routledge).

Marston, S.A. (2000), 'The social construction of scale', *Progress in Human Geography* 24 (2), 219–42.

Massey, D. (1984), *Spatial Division of Labour* (London: MacMillan).

Massey, D. (1994), *Space, Place and Gender* (Oxford: Blackwell Publishers).

Massey, D., Arango, J., Hugo, G., Kouaouci, A., Pellegrino, A. and Taylor, J.E. (1993), 'Theories of international migration: a review and appraisal', *Population and Development Review* 19 (3), 432–66.

McDowell, L. (1993), 'Space, place and gender relations', *Progress in Human Geography* 17 (2).

McLafferty, S. and Preston, V. (1992), 'Spatial Mismatch and Labour Market Segmentation for African-American and Latina Women', *Economic Geography* 68, 406–31.

Mohanty, C.T., Russo, A. and Torres, L. (eds) (1991), *Third World Women and the Politics of Feminism* (Bloomington, IN: University of Indiana Press).

Mongin, O. (1995), *Vers la troisième ville?* (Paris: Hachette).

Nagar, R., Lawson, V., McDowell, L. and Hanson, S. (2002), 'Locating Globalization: Feminist (Re)reading of the Subjects and Spaces of Globalization', *Economic Geography* 78 (3), 257–84.

Pratt, G. (1997), 'Stereotypes and ambivalence: the construction of domestic workers in Vancouver, B.C.', *Gender, Place and Culture* 4 (2), 159–77.

Pratt, G. and Yeoh, B. (2003), 'Transnational (counter)topographies', *Gender, Place and culture* 10, (2), 159–66.

Raffestin, C. (1980), *Pour une géographie du pouvoir* (Paris: Litec).

Rose, G. (1993), *Feminism and geography: the limits of geographical knowledge* (Cambridge: Polity Press).

Rutherford, B.M. and Wekerle, G.R. (1988), 'Captive rider, captive labor: spatial constraints and women's employment', *Urban Geography* 15 (3), 116–35.

Sanchez, M.J.A. (1997), 'Intra-European migrations: from North to South', paper presented at the Third European feminist research Conference, *Shifting bonds, shifting bounds: women, mobility and citizenship in Europe* (University of Coimbra, Portugal).

Sassen, S. (1999), 'Blind Spots: towards a feminist analytics of today's global economy', Paper presented at the University of Wisconsin.

Sheller, M. and Urry, J. (2006), 'The new mobilities paradigm', *Environment and Planning* 38, 207–26.

Sibley, D. (1999), 'Creating geographies of difference', in D. Massey, J. Allen and P. Sarre (eds) *Human Geography Today* (Cambridge: Polity Press), 115–28.

Silvey, R. (2004), 'Power, difference and mobility: feminist advances in migration studies', *Progress in Human Geography* 28 (4), 1–17.

Silvey, R. (2006), 'Geographies of Gender and Migration: Spatializing Social Difference', *International Migration Review* 40 (1), 64–81.

Smith, A. (1986), *The Ethnic Origins of Nations* (New York: Blackwell).

Sparke, M. (2001), 'Networking globalization: A tapestry of introductions', *Global Networks* 1 (2), 171–9.

Tarrius, A. (1994), 'Territoires circulatoires et espaces urbains', *Les Annales de la recherche urbaine*, Numéro spécial: Mobilité.

Taylor, P.J. (1999), 'The So-called World Cities: An Evidential Structure of a Literature', *Environment and Planning A* 31, 1901–4.

Thomas, C. and Villeneuve, P. (1998), 'Les navettes à Québec: genre, famille et résidence', *L'espace géographique* 3, 239–51.

Tyner, J. (2003), *Made in the Philippines: Gendered Discourses and the Making of Migrants* (New York: Routledge).

Urry J. (2000), *Sociology Beyond Societies, Mobilities for the Twenty-first Century* (New York: Routledge).

Veltz, P. (1996), *Mondialisation, Villes et Territoires* (Paris: PUF).

Villeneuve, P. and Rose, D. (1988), 'Gender and the separation of employment from home in metropolitan Montreal, 1971-1981', *Urban Geography* 9 (2).

Wittel, A. (2001), 'Towards a network sociality', *Theory, Culture and Society* 18, 31–50.

Wyly, E.K. (1998), 'Containment and mismatch: Gender differences in commuting in metropolitan labour markets', *Urban Geography* 19 (5), 395–430.

Yeoh, B.S.A., Huang, S. (1999), 'Spaces at the margin: Migrant domestic workers and the development of civil society in Singapore', *Environment and Planning A*, 31, 7, 1149–1167.

Chapter 7

The Culture of Automobility: How Interacting Drivers Relate to Legal Standards and to Each Other in Traffic

Anette Jerup Jørgensen

Introduction

This chapter is based on a theoretical and empirical enquiry into traffic culture among drivers, with a particular focus on the relationship between the law, morality, and actual behaviour. Data collection took place in the autumn of 2005 in Northern Jutland, Denmark. I used three data collection methods: filmic observation of actual driver behaviour, quantitative questionnaires for drivers, and focus group interviews with both drivers and driving instructors. The questionnaire contained questions about how the respondent believed one ought to behave in traffic interactions, how they perceived others' actions, and specifics about the respondents' own typical actions when interacting in traffic. The questionnaire was sent to a stratified selection of 200 drivers between the ages of 18 and 73 chosen from the Central Register of Motor Vehicles in Denmark, with half of the respondents being men and half women. The results of the questionnaire provided an impression of driver's attitudes and actions, and were used to select participants for two follow-up focus group interviews, with eight drivers in each group. The drivers were selected in a way that ensured that the participants were as different as possible. To kick-start the group discussions, I used filmic observation of traffic interactions. By discussing specific actions, it was possible to capture variations in the normative praxis and the justifications drivers used for their actual behaviour.

It is in this context that the dimension of gender is documented due to the fact that gendered implications appear to be significant.[1] The purpose of this chapter is to elucidate the gendered implications of traffic culture in a theoretical discussion about how drivers relate to one another and to the law. This also includes discussions about how drivers relate to the risks associated with the culture of automobility[2]

1 When using the term gender, I refer to the cultural assumptions and practices that govern the social construction of men, women and their social relations.

2 In this chapter, the term culture refers to the socially transferred or learned ways of living and habits to which symbolic patterns, norms and rules belong. It is a combination of ideas, praxis, material and symbolic scopes of meaning, which results from individuals' social activities within different groups or societies. A culture functions as a producer of meaning but also outlines rules for social behaviour and action (Ekegren, 1998, 140–41; Frönes, 2001, 13).

due to the fact that objective risks of traffic accidents are used as a moral yardstick among drivers to determine which actions are right or wrong. However, there is not always an association between the objective risks and the moral risks, because the objective risks can be dramatised or played down from a moral perspective based on what is perceived to be right or wrong. The objective of this chapter is to contribute to improved understandings of the matter of moral and legal standards in car traffic interactions, and to grasp in what way drivers' actions and their way of thinking are gendered. Given that the gendered implications associated with this sociological enquiry primarily emerged out of the empirical data, I will emphasise some of it in my reflections in order to strengthen the theoretical view. This is mainly done because the existing sociological research on how drivers relate to law and to one another in traffic has been investigated in only a limited way within the automobility paradigm.

The theoretical framework for the discussion

In order to really grasp how drivers relate to the law and to each other in traffic it is necessary to include theories at two levels. Macro level theories describe how structures and processes of change influence norms and actions among individuals and groups in late modern societies. This more than a unidirectional influence because perceptions and actions also influence structures and thereby initiate processes of change. From this perspective, overall societal conditions of high modernity influence the way drivers relate to one another in traffic.

In the interaction process, drivers internalise norms and choose whether or not to conform to established norms. Hence, it is also necessary to include micro level theory in order to fully understand how, and in what way, drivers relate to one another and how the relations are gendered. To give the reader an overview of what is to come, I will briefly introduce the theoretical terms that are the subject of this discussion.

Theories about high modernity can be used to frame the discussion because high modernity proves to be the context within which moral and legal standards are produced and the context within which drivers relate to one another. The separation of time and space and associated mobility are essential features of late modern societies. In the wealth of possibilities that result from the reorganisation of time and space and society's associated geographical mobility, automobility is established as a risky practice from a material, cultural, and social standpoint. Automobility and its risks are seen as a product of the technological development that has taken place in late modern societies at the same time as individualisation is on the increase. This growth has made moral and legal standards increasingly important in maintaining social order. However, the validity of norms depends on communicative contracts, which in principle ought to be agreed upon by everyone, but realistically can be

---

In this sense culture is not perceived as stable and unambiguously confined but as dynamic and changeable over time, so that the dominating symbolic patterns in society can be influenced by other cultural patterns and interpretations (Frönes, 2001, 13–14).

affected by existing power relations, such as differences in male and female drivers' ability to establish moral standards in a traffic culture.

High modernity

Although the concept and consequences of modernity are quite complex, sociological thinkers who are concerned with the issue agree that the western worldview changed with the development of industrialisation and rationalisation (Beck, 1997, 16; Ziehe, 1989, 16, 155). According to Giddens, modernity refers to the social life and organisation structures that have been under development in Europe since the 17th century, and since then have reached more or less worldwide significance (Giddens, 1994, 9). The changes in life and organisational structures that have been generated by modernity mean that we, in unprecedented ways, have removed ourselves from traditional types of order because of the dynamics of modern institutions and the way they undermine traditional habits and practices (Giddens, 1994, 12, 39; Giddens, 1996, 9). However, this does not entail the disappearance of communities, but merely their transformation. Social actions in modernity are emancipated from their former local constraints and social relations are released from local time cohesion, which entails the separation of time and space (Juul, 2002, 90). This is of utmost importance for the development of other forms of social life and the establishment of new kinds of communities. It is an assumption that I will develop in the following section with direct reference to automobility and the social relations it involves.

Automobility

According to Giddens, the dynamics of modernity is driven by a reorganisation of time and space, so that social relations are disembedded from their local constraints and reconstructed across time and space (Giddens, 1994, 26; Giddens, 1996, 11). At the same time, the need for geographical mobility is drastically increased. For this reason individuals have to travel over longer distances in order to engage in their social lives. Intense geographical mobility has been characteristic of modern society since the mid-1800s. This, in turn, has influenced the development of the modern city as expanding transportation innovations have further increased geographical mobility (Hjorthol and Lian, 2004, 1). The development of motorised vehicles can be seen as a mutually dependent development, where the increased need for geographical mobility generates the need for motorised vehicles, which again generates the possibility and need for further automobility.[3]

---

3 This could easily be mistaken for technological determination. So far, automobility has proven itself to be an automotive development from a structural, cultural, and social perspective. However, at this point the increasing number of cars has led to intense road congestion in large cities all over the western world. It makes politicians as well as populations aware of the downside of the automotive development of automobility. In response, many western countries are initiating efforts to downplay the impact of automobility and efforts are being made to shape alternatives to automobility. Hence, the relation between the development of motorised vehicles and the increased need for geographical mobility may not be seen as technologically determined.

Interestingly, the car is often described as a vehicle of freedom – one can go wherever and whenever one desires without considering bus schedules or railway timetables. In other words, it entails freedom from some institutions or regularities and in this sense from collectivities. Hence, in a car-dependent society, the car is said to bring the mechanical liberation of the individual (Hagman, 2003, 3). It is an example of how detraditionalised ways of life shape a new relationship between the individual and society, where any social influence on individual action is experienced only indirectly and to a limited extent (Beck, 1997, 120). Automobility is not only socially but also materially restricted. It is only possible to use the automobile according to the way it is engineered. The automobile cannot exceed its physical/material construction and, thus, drivers are restricted by such things as the car's maximum speed and built-in steering limitations (Otnes, 1994, 12). Social constraints reveal themselves when the mobile individual is re-embedded in another social structure – a standardised automotive way of living. This is most obvious in the case of gridlock. If automobility gives an individual freedom, it is a freedom that only exists until the driver is trapped in a traffic jam, which illustrates how individualisation is related to institutionalisation, or how the freedom of the individual is associated with new institutions surrounding the driver (Beckmann, 2001, 48–49). Even though this may be difficult for the individual driver to comprehend, it is nonetheless the growth and the increasing awareness of this antagonistic condition that establishes the possibility for emerging social and cultural communities (Beck, 1997, 121–2). Among other factors, the opportunity to develop new social and cultural communities is conditioned by drivers' awareness of traffic as a social arena, a social practice where one's own automobility is intertwined with that of others. One can say that it is a forced need for sociality, because when in traffic, the individual driver is not actually seeking the company of others. The awareness of automobility as a socially institutionalised praxis is necessary in order to counteract a driver's tendency to try to achieve more automobility than the driver in front of them, which only entails enormous queues in traffic. When drivers try to intensify their automobility in order to save time when travelling in space, everyone is trapped in gridlock with the tragicomic result that everyone becomes immobile. Consequently, what Giddens (1996) believes to be unrestricted mobility in time and space is really an illusion; boundless individual mobility does not exist today. Still, drivers may believe it to exist and try to achieve free individual automovement. How drivers relate to one another and crave free auto-movement is also contingent upon other factors, such as the gendered distribution of cars and the gendered differences in relation to the labour market.

Traditionally, the automobilisation of family life did not only bring the newest and most expensive car models first to the male heads of families, while the women had to settle for second-hand models or smaller cars, but also led to the uneven gendering of time and space. While working men became enmeshed in the stress of everyday commuter traffic into and out of urban centres, suburban housewives had to juggle family time around multiple, often conflicting, schedules of mobility epitomised by the school run and 'mom-as-chauffeur' (Urry, 2006). The uneven gendering of time and space has also influenced and established a gendered perception of automobility. Every driver now and then runs into congestion or experiences other ways in which the presence of others limits their individual automobility. In the case of two lanes narrowing to one, where one driver is compelled to stop in order for the flow of traffic to continue, both male and female participants in my study found that male drivers became more frustrated than the female drivers when they had to stop for another driver to pass. Most of the male participants would rather push forward to keep moving, while most of the female drivers seem to have little concern about bringing their own car to a standstill in order for the other driver to pass. Hence, in such cases, women tend to be more tolerant towards delays than men. Some of the women in my enquiry described this difference by saying that if it is so important for *him* to attain free mobility, the women would rather be constrained in their own automobility than participate in a competition for free individual movement.

When in traffic, women also tend to be more concerned about children than men. The forbearance of women on the move could then be caused by the historically uneven gendering of time and space, because women traditionally have been left with slower cars than men. And because of their role as mom-as-chauffeur, women become subordinated in terms of free mobility because their mobility patterns to a greater extent than the male commuter are conditioned by the transportation needs of the family. The subordination of women in traffic influences how automobility is perceived and how drivers relate to each other in traffic. This I will clarify later in the chapter as regards to the negotiations[4] about what is perceived as morally acceptable in traffic.

Concepts of risk

Firstly, I would like to take as my starting point the assumption that free individual mobility is really an illusion, because of moral and physical restraints. One of the basic moral norms in traffic is that drivers who disregard the risks related to driving have to deal with severe accidents, repair costs and damage, and the sanctions that the legal system has to offer (Otnes, 1994, 12). This basic moral norm in traffic illustrates how the responsibility is placed on drivers who act immorally, even though risks now form a globally hazardous condition for all of mankind. In high modernity, a new risk fate is ascribed to the individual, who cannot escape this no matter how much they try. From Beck's point of view, modern risks differ fundamentally from the risks that were common in earlier stages of industrialisation because the threat now is global and the causes human-inflicted (Beck, 1997, 10, 30–31). Hence, even though risks are unavoidable for everyone, the responsibility for risks is ascribed to the individual (Beck, 1997). The responsibility for risks in traffic is morally placed on drivers who challenge social and technological constraints in spite of the fact that mass motoring is a risky social praxis that in itself generates automobile risks.

It is difficult to decide whether risks have become more widespread than previously or if it is only our perception of them that has changed. However, one thing is certain

---

4 In this sense the term negotiation is used to describe the social activity that takes place both in and out of traffic interaction, where drivers discuss what are morally acceptable actions in practice. The negotiations can be harmonious as well as marked by conflict. It depends on whether or not it is possible to reach a communicatively moral agreement.

– risks can be dramatised, reduced, increased or downplayed depending on existing knowledge (Beck, 1997, 32, 74). Risks do not only involve the probability of an accident occurring, but also a set of values determining what morally acceptable actions are (Douglas, 1985, 31, 60). This point is essential in the following discussion of how drivers relate to each other, because the gender of the driver influences the perception of whether or not objective risks should be downplayed or emphasised. Ultimately this also has implications for social negotiations about what is morally correct behaviour in traffic interactions and thus for the way driver relations are gendered.

According to Douglas, the modern concept of risk is aimed to protect every single individual in the community from the abuse of others and so serves to maintain an individualised culture (Douglas, 1992, 28). The important question, however, is how drivers determine which actions are morally risky, because there is no given consistency between the objective risk and the moral judgement of the risk. As opposed to the perception of risk that Beck advocates, defining moral risks is not only a matter of determining the risk of road accidents. It may also involve the risk of immobility, the risk of wasting time, as well as the risk of being late to work. All these different kinds of risks are taken into consideration when we decide what is morally risky. The dominant moral definitions of risks set the boundaries for the freedom of action for the individual driver. As long as one's actions are not considered to be morally risky to another's health, they are accepted.

Women are often brought up to be more concerned about risks than men (Douglas, 1985, 70), which causes them to be more concerned about the objective risk of road accidents than men. My enquiry indicates that male drivers tend to downplay the importance of objective risks if the objective risks restrain their individual movement in specific situations. Often the risk of losing time and being late for work are more valued than the risk of a traffic accident. Hence, the valuing of risks is largely gendered.

It is not only the objective and moral risks that constrain the movement of drivers. Even though automobility causes expectations about free individual movement, everyone is more or less aware that the way to improve and secure one's own condition depends upon regulation of the actions of others so that the way is paved for fulfilling one's own actions and interest. This also implies duties and obligations towards others (Tonboe, 1997, 74). In a traffic culture, both legal and moral norms determine the rights and obligations drivers have towards each other when interacting in traffic.

Law and morality

Traffic behaviour is highly regulated by law, with the purpose of integrating road users according to common rules to maintain social order. The comprehensive regulation by law can be seen as a means to inhibit ruthlessness and self-interested behaviour in traffic (Bjørnskau, 1993, 48). According to Habermas, legal standards define the same liberties for every individual who is the carrier of rights, and within these rights drivers can express their own free will. Thus, modern law is well suited for social integration in differentiated societies (Habermas, 1975, 95; Habermas, 1996, 82–4). If normatively regulated everyday communication fails, however, or the normal co-ordination of actions is dysfunctional, the law as well as morality are structured to handle open conflicts by which a higher level of consensus can be reached through reasoning. The similarity between legal and moral standards lies in the fact that both refer to how interpersonal relations can be ordered legitimately, how actions can be co-ordinated through justifiable norms, and how actions in conflict with these norms can be resolved by the use of consensus about intersubjectively recognised normative rules and principles. Even though moral and legal standards refer to the same issue, legal standards are only legitimate if they do not conflict with basic moral principles (Habermas, 1996, 106; Habermas, 1997, 353). At first glance, men and women enjoy the same legal rights and obligations in traffic, which instantaneously establishes gendered equality. There is actually gendered equality when the legal standards are perceived as morally legitimate and guide drivers' behaviour in traffic interactions. Unfortunately for female drivers, legal standards do not necessarily enjoy moral recognition and therefore drivers negotiate which norms should regulate traffic behaviour.

The negotiation of what is right or wrong is very common in all arenas of high modernity. Intersubjectively recognised norms are constantly changing in view of the reflexivity of individuals. As a consequence, respect for legal standards is no longer an automatic given – it is only given when individuals think that respect has been earned (Andersen, 1998, 307). In principle, drivers can have the opinion that one should play by legal standards, but in specific situations individuals are most likely to act reflexively and negotiate legal standards morally (Andersen, 1998, 395). In case of a change in morality due to the reflexive orientation of individuals, the legitimacy of legal standards can be reduced, which can prove to be problematic for female drivers because the negotiations do not always ensure that their point of view is considered. For instance, in my enquiry male and female drivers discussed whether or not one should be allowed to turn to the right at a traffic light intersection if nobody else is around, even if the traffic light is red and even if it is illegal to do so in Denmark. The male drivers agreed that it is a legitimate action and that it would be appropriate to change the established rule of law in order to make the action legally legitimate. Most of the female drivers disagreed. To a greater extent than the male drivers, they believed that the established law should be maintained. They were more sceptical about negotiations, arguing that there are also children in traffic and it is important to have well-defined rules that are easy for everyone to understand, including children. Women perceived consideration for children as more important, while men were more concerned about how they could improve their constant movement in autospace. The fact that women tended to be more law-abiding in traffic than men was not limited to traffic interactions. As mentioned earlier, this is also true in other regulated arenas of society. Other studies demonstrate that men generally qualify and negotiate the restrictions given by law to a greater extent than women (Andersen, 1998).

Apart from the gendered differences that accompany the negotiations of law, the lack of moral recognition of the law can in itself prove to be a serious legitimacy problem. It may also prove to be problematic if drivers only relate to legal standards and thus disregard consideration for other drivers. Habermas emphasises that individuals' private legal autonomy can be conceived as the negative freedom given by the individual who upholds the law and thus who does not have to justify their actions towards others, as long as those actions are legal. This may undermine communicative action and induce drivers to act egoistically (Habermas, 1996, 120). When the actions of a driver are perceived to be egoistically motivated, the reactions from other drivers appear to be particularly gendered. This I will discuss further in the following sections, where I present some coping strategies in traffic interactions and detail how drivers respond if these coping strategies are violated by others.

Coping strategies in traffic interactions

Initially, it is necessary to clarify my perspective in relation to one of the most common assumptions of infringements of law in traffic interactions. Often drivers who disobey the law in traffic are conceived as self-centred people with little regard for others. As I will argue here, this understanding of infringements has to be reconsidered and revised. According to Habermas, there are two different types of actions: communicative and strategic. Strategic actions apply to individuals who pursue their aim through responding to and influencing the decisions of others (Habermas, 1997, 126). In contrast to strategic actions, communicative actions are embedded in norms and values and can be perceived as right or wrong (Frönes, 2001, 38). Often traffic interactions are conceived as solely strategically determined. However, acting strategically, or looking after one's own interests, is not necessarily irreconcilable with communicative action. My enquiry indicates that as long as drivers pursue their own interests, taking due account of what is commonly morally recognised as proper manners, their actions are perceived as morally acceptable. This may be seen in relation to the moral risk concept, which establishes that actions are perceived as legitimate as long as they do not pose risks to other drivers.

To some extent, individuals need autonomy and room to manoeuvre in order to be able to survive and shape their own lives in a detraditionalised and individualised society. But such autonomy also entails cohesion and reciprocity. Modern man is trying to combine self-realisation with being for others; hence individualism, understood as self-realisation, is not inconsistent with altruism (Giddens, 1994, 13). In order for drivers to achieve their individual objectives within moral boundaries in traffic, social coping strategies are formed, helping drivers to obtain their objectives without them being at the expense of others. These coping strategies work as a supplement to legal standards. This is a way of maintaining social order and avoiding the risks associated with traffic interactions.

One of the most important coping strategies in traffic according to Goffman (2004, 167) is probably 'the first-arrived principle', which establishes that a driver has the right to go immediately after the driver in front and immediately before the driver at the rear. A related principle is 'the principle of equality', which involves a queue arrangement where the driver who arrives first has the right to emerge first.[5] This creates a time arrangement that completely blocks the influence from otherwise existing social relations and differences in the social status of drivers (Goffman, 2004, 168). However, there can be situations where queue rules are suspended due to extraordinary circumstances. An example is the case of a situation where two drivers arrive simultaneously where two lanes narrow down to one, and where there is a queue in both lanes. Here drivers use more far-reaching principles as a kind of noblesse oblige, where the driver with most social status or power allows the other to go first, much as a protector would act towards the protected (Goffman, 2004, 318). Both male and female drivers agree that in such cases one ought to let elderly drivers, tourists, and learner drivers in, even though it disregards the first-arrived principle. It is commonly recognised that if drivers with these characteristics are not allowed to shift their lane it is an egoistic act of the driver not letting the person in. In other cases, if a driver with characteristics other than those mentioned above tries to ignore the moral queue rule, it is considered to be egoistic behaviour – and most often punished. How this is done will be clarified later in the chapter. Even though male and female drivers agree upon informal moral queue standards, women tend to give way to others in line to a greater extent than male drivers, probably because men are more oriented towards individual automobility than women. How is it then possible to form and maintain these social coping strategies in traffic, when the interaction is fluid and the communication conditions are said to be constrained?

Automobile communication

According to Habermas (1997, 124), communicative action determines what is morally acceptable and what is not. In order for communicative action to function properly it requires interaction between two subjects capable of speech and action to enter into an interpersonal relationship with verbal or nonverbal means. The individuals seek a common understanding about the acting situation in order to coordinate their plans for action and hence also their actions. There is a need for justification of norms in late modernity, and as a consequence the situations are frequently open in such a way that the communicative agreements are only partly given in advance. For this reason the agreement is in part formed in the interaction process (Frönes, 2001, 64). The outcome is that drivers negotiate with one another in order to determine what is right or wrong in a moral sense.

The typical way of seeing the automobile is as a device that restricts and constrains communication since the human body is encapsulated in a metal cocoon. Spoken words, eye contact, and facial expressions are normally missing (Featherstone, 2004, 12). From this perspective automobility creates distance between people, where the only thing people see of one another is a head and shoulders with glass around it. According to Otnes (1994, 26–7), this gives drivers a fixed focus world-view that results in difficulties in communicating with other drivers. It also explains

5 The two fundamental principles are closely interrelated and in practice they mostly occur simultaneously. Thus it may be most lucrative to understand the principles as analytical constructions, which in reality often fuse together.

why drivers often conceive others egoistically. Because drivers are not able to communicate their worries properly, a high amount of ambiguity is introduced in driver interaction (Thrift, 2004, 47–8). Thus communicative action in traffic is impossible to establish.

However, from another perspective, automobility is seen as a new communication platform with a complex set of possibilities. There are ways of communicating through the windscreen, windows, and mirrors that entail interactions and ways of presenting the autoself in fluid interaction. In interacting with different reference groups, the driver has two bodies; the one that is the automobile and the other that is the driver's own body (Featherstone, 2004, 8, 12). According to Urry (2003, 7) the automobile becomes an extension of the driver's own body in such a way that the driver can feel the contours and shape of the automobile. Similarly, Thomsen (2001) has described how drivers communicate symbolically due to the fact that the automobile does not only refer to the self-image of the driver, but also to a common understanding of the individual and the community. The automobile expresses social and economic status, along with the kind of behaviour that is expected of the driver. Even though the communication possibilities are constrained, other ways of communicating intentions in traffic exist, including the use of the driver's own body as well as the body of the automobile.

Drivers, regardless of gender, are very capable of reading and understanding each others' signals in traffic, even though it is seldom possible to use the spoken word. There are some commonly recognised communication strategies in these situations. If a driver at the rear drives too close, both men and women may gently touch the brake in order to inform the driver behind that the driver is driving to close. Another method is to use the vehicular hazard warning signal flasher (Jorgensen, 2007). However, automobile communication strategies are often gendered with far-reaching consequences. Even though women use active communication strategies, male drivers tend to communicate actively to a greater extent than women – especially when other drivers violate moral standards in traffic. For instance, male drivers tend to express their opinion by flashing headlights if someone overtakes in a dangerous manner. In the case of another driver trying to violate the first-arrived principle, they tend simply to ignore the driver by looking the other way and continue to drive ahead without giving way to the other driver. Both are obvious examples of automobile communication. Women, on the other hand, tend to react more passively than male drivers when others violate moral standards in traffic. If a (male) driver tries to jump the queue and elbow past a female driver, she is most likely to tell the driver off in the privacy of her own car but still give way to the driver.

Even though automobile communication is especially active in the case of a violation of norms, it is also employed in the recognition of courtesy and politeness. For instance, if a driver lets another in from a by-road, recognition can take the form of nodding or putting up the arm for a little wave to signal gratitude. However, the signals for courtesy are not particularly gendered. Despite this, gendered differences in automobile communication strategies have far-reaching implications because they also entail gendered differentiation when it comes to the ability to establish and maintain moral norms in traffic. This can prove critical for the opportunities for communicative action because it emphasises that everyone must have equal access to determine moral norms (Habermas, 1975, 116–18). When men tend to communicate more actively in traffic interaction than women, they are more able to establish and maintain moral standards. This can sustain the traditional order in traffic – an order that is still dominated by masculinity (Andréasson, 1994, 138). It emerges as a severe problem in cases where men and women disagree on which standards should regulate interaction. Male and female drivers have different views of highway speed, as one example. Male drivers tend to view legal speed limits as optional. For this reason they think that it is acceptable to exceed legal speeds limits in the fast lane, as long as this does not pose a risk to anyone. As most women tend to be law-abiding, they respect speed limits when overtaking, even though it results in tailbacks. This creates conflicts between men and women on the road because it involves relative immobility (especially for men who like to drive fast). This is a conflict that is conceived of as male-dominated by both men and women, because male drivers tend to sanction the law-abiding behaviour of others. They do so by driving up closely behind in order to make the driver (typically a woman) move faster so that the male driver is not frustrated in his aspiration for maximum individual automobility. Thus male drivers more actively use automobile communication strategies to give them greater impact in setting moral standards in traffic. The differences in ability to set moral standards may be caused by the fact that often the road has been perceived primarily as a masculine territory that genders the experience of travelling as a masculine identity project, which constructs women as socio spatially other; as atypical travellers (Enevold, 2000; 403–6). This gives the impression of a male-dominated traffic culture with gendered differences in the power to define how automovement is to be perceived, what actions ought to be perceived as legitimate, and also what is perceived as morally risky.

Summary

The objective of this chapter has been to elucidate how drivers relate to one another and to legal regulation when interacting in traffic. Within this context, gendered distinctions have been investigated. Gender appears to play an important part in the culture of automobility. Male and female drivers have different approaches to automobility and its risks in regards to the negotiations of what is morally acceptable behaviour, and also to automobile communication in traffic interaction. Socially constructed gendered approaches are of importance in how drivers relate to each other in traffic.

Comprehensive legal regulation is intended to establish equal rights in a highly differentiated society. But because of the contingency between law and morality in an individualised society, where drivers act reflexively, legal standards are constantly being subjected to gendered negotiations about what is morally acceptable when interacting in traffic. As women tend to be more oriented toward the mobility of collectivities and more law-abiding, men are more concerned with achieving free individual automovement and negotiating the reasonableness of legal standards. In the negotiations about which moral standards should regulate practices, objective risks are emphasised or played down in relation to what is perceived as morally acceptable. While women tend to be more concerned with objective risks and are more law-abiding, men tend to perceive free individual automovement as more important than the consideration of objective risks. This creates conflicts in the interactions among drivers. Often male drivers come out better in these conflicts than women because of the gendered differences in the use of nonverbal automobile communication. In the case of a gendered disagreement about what is the acceptable approach to traffic interaction, male drivers tend to communicate more actively in traffic, punishing the driver who does not live up to moral standards. It seems that female drivers have greater difficulties getting their way when it comes to the definition of moral standards, partly because their communication strategies in traffic tend to be more passive.

The different gendered understandings of how people should relate to one another in traffic contribute to the maintenance of the masculine-dominated culture of automobility.

References

Andersen, J.G. (1998), *Borgerne og lovene* (*The Citizens and the Laws*) (Aarhus: The Research Unit of the Rockwool Foundation and Aarhus Publisher).

Andréasson, H. (1994), 'Ensam i bilen' ('Alone in the Car'), in B. Bursell and A. Rosengren (ed.), *Drömmen om bilen* (*The Dream of the Car)* (Uddevalla: Nordic Museum).

Barker, C. (2000), *Cultural Studies - Theory and Practice* (London: Sage Publications).

Beck, U. (1997), *Risikosamfundet (Risk Society)* (Copenhagen: Hans Reitzels Forlag).

Beckmann, J. (2001), *Risky Mobility – The filtering of automobility's unintended consequences* (Institute of Sociology: Copenhagen University).

Bjørnskau, T. (1993), *Spillteori, Trafikk og Ulykker: En Teori om Interaktion i Trafikken (Game Theory, Traffic and Accidents: A Theory about Interaction in Traffic)* (Oslo: Transport Economic Institute, Oslo University).

Douglas, M. (1985), *Risk Acceptability According to the Social Sciences* (London: Routledge & Kegan Paul).

Douglas, M. (1992), *Risk and Blame - Essays in cultural theory* (London and New York: Routledge).

Ekegren, P. (1998), 'Kultur' ('Culture'), in Heine Andersen et al. (eds.): *Leksikon i sociologi (Sociological Encyclopaedia)* (The University Publisher A/S & Academic Publisher A/S).

Enevold, J. (2000), 'Men and Women on the move – Dramas on the road', *European Journal of Cultural Studies* 3 (3), s. 403–20 (London: Sage Publications).

Featherstone, M. (2004), 'Introduction', in V. Bell et al. (eds): *Theory, Culture and Society. Special Issue on: Automobilities* (4/5), August-October.

Frönes, L. (2001), *Handling, kultur og mening (Action Culture, and Meaning)* (Bergen: Specialist Books Publisher).

Giddens, A. (1994), *Modernitetens konsekvenser. (The Consequences of Modernity)* (Copenhagen: Hans Reitzels Publisher).

Giddens, A. (1996), *Modernitet og selvidentitet. (Modernity and Self-Identity. Self and Society in the Late Modern Age)* (Copenhagen: Hans Reitzels Publisher).

Goffman, E. (2004), *Social samhandling og mikrosociologi. (Social Interaction and Micro Sociology)* (Copenhagen: Hans Reitzels Publisher).

Habermas, J. (1975), *Legitimationsproblemer i senkapitalismen. (Crisis of Legitimacy in Late Capitalism)* (Copenhagen: Fremad).

Habermas, J. (1996), *Between Facts and Norms* (Cambridge: Polity Press).

Habermas, J. (1997), *Teorien om den kommunikative handlen (The Theory of Communicative Action)* (Aalborg: Aalborg University Publisher).

Hagman, O. (2003), *Mobilizing meanings of mobility: Car users' constructions of the goods and bads of car use*, Transportation Research Part D 8 (2003) 1-9. (Göteborg: Science and Technology Studies).

Hjorthol, R. and Lian, J.I. (2004), *Sammendrag: Samfunnsmessige trender – betydning for mobilitet og transport i storbysamfunnet. (Summary: Social Trends – the Significance for Mobility and Transport in Urban Society)* (Oslo: TÖI-report 718/2004).

Jorgensen, A.J. (2007), *Trafikkultur – retten, moralen og den faktiske adfærd. (Traffic Culture - the Law the Moral, and the Actual Behaviour)* (Aalborg: Aalborg University Publisher).

Juul, S. (2002), *Modernitet, velfærd og solidaritet (Modernity, Welfare and Solidarity)* (Copenhagen: Hans Reitzels Publisher).

Otnes, P. (1994), *Can we Support ourselves by Driving to each other? Collective and Private Transportation: How the Automobile has Affected Us, What Collective Transportation Does Differently, and Why. A Chreseological Suite in Eight Movements* (Oslo: Institute of Sociology).

Thomsen, T.U. (2001), *Persontransportens betydning for individet i et identitetsperspektiv – med fokus på transportmiddelvalg. (The Meaning of Passenger Transport for the Individual in an Identity Perspective focused on Choice of Means of Transportation)* (Aarhus: Institute for Market Economy/ Center for Social Scientific Environmental Research).

Thrift, N. (2004), 'Driving in the city', in I.V, in Bell et al., *Theory, Culture and Society. Special Issue on: Automobilities* (4/5), August-October (Oxford).

Tonboe, J. (1997), 'Sociologisk sans, moralsk forandring og princippet om kontrafaktisk fortolkning' ('Sociological Sense, Moral Change and the Principle of Counter Factual Interpretation'), in *Dansk Sociologi (Danish Sociology)* (2/8) (Copenhagen: The Danish Sociologist Association).

Urry, J. (2003), *Inhabiting the car* (Department of Sociology: Lancaster University).

Urry, J. (2006), 'Inhabiting the car', *The Sociological Review* 54, 17–31.

Ziehe, T. (1989), *Ambivalenser og mangfoldighed: En artikelsamling om ungdom, skole, æstetik og kultur (Ambivalence and Diversity: A Joint Publication about Youth, School, Aesthetics and Culture)* (Copenhagen: Politisk Revy).

PART 2
How and Why are Mobilities Gendered?

Chapter 8

Gender Still Matters: Mobility Aspirations among European Scientists Working Abroad

Elisabeth Scheibelhofer

Introduction

This chapter focuses on the gendered mobilities of men and women crossing the international borders of developed countries. The considerations presented here are based on an empirical qualitative study including qualitative interviews with 21 Austrian scientists. The research explored the mobility aspirations of these researchers who had been working in the USA on a medium term or longer basis at the time they were interviewed. The analysis (as described later in this chapter) led to the construction of three ideal types of mobility: Migration, short-term research stays, as well as transnational modes of mobility.[1] As current research indicates, transnational mobility aspirations have been neglected for a long time, especially with a focus on the impact of gender.

With the example of one case study reflecting a transnational mobility pattern, the chapter will focus on gendered forms of mobility and demonstrate that even such socio-economically privileged migrants share characteristics with migrants from developing countries: although mobile scientists from Western European countries are in general less restricted by visa regulations, economic hardship or discriminatory practices, the interviewees had to deal with various tensions resulting from their wish to lead self-determined lives on the one hand and the restrictions of a mobile lifestyle on the other. These limitations are often based on gendered ascriptions and orientations, as in the case of a transnationally oriented scientist of Austrian origin. Although she has established her professional and private life in the USA and in Austria, it is clear to her that once her parents need her help and care, she will leave and settle in her parents' rural residence in Austria. Still, she refuses to dwell on the consequences – as both her work and her relationship could very well be put on hold once this situation occurs. Thus, comparing the personal relations of mobile women from a developed country with the pitfalls associated with the mobile lives of women from developing countries reveals differences as well as similarities: the family's economic well-being, or personal or political freedom, is not the most pressing motivation behind the desire to lead a mobile life, as is frequently the case in

1 For a detailed report of the results of the study see Scheibelhofer, 2003; 2005.

transnational mobility from developing countries. However, the gendered difference in caring for social relations – a role that is still ascribed to women – can play a key determining factor when looking at physically highly mobile women from Western Europe, as this case study demonstrates.

The missing argument: Highly qualified women and mobility

The case that Eleonore Kofman (2000; 2004) makes in her contributions regarding gendered global migrations also holds true for the issue developed in this chapter. Kofman describes how the globalisation of migration has also brought about an increasingly diversified and stratified form of migration. Although approaches such as transnationalism (Basch et al., 1994; Levitt et al., 2003; Pries, 2002) or alternative circuits of globalisation already reflect the complex character of migration under present day circumstances, they have not adequately addressed the diverse forms of gendered stratification produced by migration, both in migrants' countries of origin and in their places of settlement (see Kofman, 2004, 644 f.). I agree with Kofman in arguing that much of the scholarly attention has been focused on socio-economically disadvantaged female migrants working in private households or in the sex industry (Kofman, 1999, 53f.). Such a focus is also well chosen since it contributes to a better understanding of the precarious living circumstances among migrant women in Western European countries without omitting their own agency in difficult everyday situations.

Yet, this important branch of research needs to be complemented by another approach that will help to elucidate the actual social reality of gendered stratification, as connected to and caused by mobility and migration. For such an endeavour, it is also essential to conduct research that concentrates on skilled mobile women and female migrants working in developed countries. As Kofman (2000) points out, skilled female migration is not a mere footnote to migration, as we are currently witnessing an increasing process of feminisation in the welfare sectors nourished by migrant women in many developed countries (Wichterich, 1999). Thus, new lines of inclusion and exclusion are opening up between those skilled and those unskilled in developed countries (Kofman, 2004), and also – as has been neglected up to now in this discussion – between skilled male and female migrants from developed countries.

One important arena in which such gender differences emerge between mobile men and women from developed countries is the degree of mobility among highly skilled individuals such as scientists. For scientists, international mobility is an important steppingstone along their career paths. Especially in Europe, international experiences in teaching and research are valued highly when institutions make staffing decisions. In contrast to the business sector, however, structural aid is scarce in terms of international mobility. While staff mobility is often promoted in companies through institutionalised procedures, scientists have the burden of organising their departure from their institution as well as their stay abroad. Additionally, mobility programs in academia are seldom accompanied by financial support for the family

– a fact that is particularly problematic for female scientists who also have to care for children (ETAN, 2000).

One might argue that the research gap left by feminist migration scholars on the topic of highly skilled women could be closed by considering investigations that focus on scientists' international mobility. This research has been quite well established since the 1960s and is mostly based on the assumption that highly skilled scientific migration (HSM) represents a unidirectional flow of individuals and of knowledge. This leads to the assumption that migrants' countries of origin are losing 'human capital', while their countries of settlement are increasing their share of highly skilled inhabitants (see Lowell, 2003). The circumstances and consequences of scientists' mobility have thus been mostly generally discussed under the well-known heading of 'brain drain'. Until the late 1980s, nationally oriented political measures attempted to counteract or regulate highly qualified citizens' emigration. One strategy was to impose specific punitive taxes on these people when they emigrated in order to compensate for anticipated negative national economic implications (see Hillmann and Rudolph, 1996). Such measures have been used vigorously in countries especially affected by the exodus of highly qualified persons, such as South Africa and Columbia. These measures to counteract emigration among the highly skilled, including scientists, have been evaluated, changed and re-evaluated. Still, the studies showed their overall impact on emigration to be minimal (ibid.). So why did these measures fail? It seems very likely that the understanding of the social processes underlying these measures was incorrect and that therefore the political activities were in vain. As described by Meyer and Brown (1999), understanding of this kind of emigration was exclusively based – at least until the late 1980s – on assumptions informed by human capital theories. According to these conceptions, qualified individuals' relevant skills result from investments made in their education.

Another approach – more convincing from a sociological point of view – is that a human capital perspective would only capture one part of an individual's relevant body of knowledge (Regets, 2003), as monetary values are measured but learning processes are ignored. Yet, learning is a complex and time-consuming socio-cognitive process. Complementary aspects of codified knowledge and tacit knowledge (see Polanyi, 1962) are brought to the fore against the background of these assumptions, together with a heightened sensibility towards the processes of knowledge production (see Gibbons et al., 1994) and innovation in the digital, so-called information society (see Brotchie et al., 1987). Therefore, knowledge (and especially its tacit part) is seen as socially and spatially embedded in specific interactions such that it can only be mediated within social processes, as Lundvall and Borrás (1999) explain:

> This implies that it [tacit knowledge] cannot be sold and bought in the marketplace and that its transfer is extremely sensitive to social context. (Lundval and Borrás, 1999, 32)

Tacit knowledge is also linked to the capability to act and to the capacity to integrate further knowledge. This concept of knowledge perceives the resources of highly qualified individuals differently from a human capital-inspired approach. The collective aspect of knowledge production then becomes the centre of attention. Cognitive networks of individuals and institutions forming a scientific community are pivotal, as specialisation is based on local conditions for research. Such environments are not easily duplicated. Thus, Powell and Smith-Doerr (1994) and other network theorists emphasise the close interrelationship between individual competencies and the units to which a person is connected (Callon, 1986; Callon and Latour, 1981). These findings have a serious impact in considering researchers' international mobility. The network perspective can make diverse forms of cooperation visible that might not be explained with a human capital approach, which instead would focus on the market and hierarchy as explanatory variables (see Schulz-Schaeffer, 2000). Studies on the meaning of social networks for men and women, and their differing status in such networks, also suggest that female scientists frequently engage in less powerful networks than their male colleagues (see Bozeman and Corley, 2004). Another critique of brain drain research is that an economic orientation fails to look at important factors shaping mobility decisions, such as household strategies (see also Scheibelhofer, 2005). Helen Stalford (2005, 361) notes in this respect that 'the impact of moving on personal and family lives is an equally important dynamic shaping mobility decisions and aspirations'.

Perceiving knowledge as embedded in social relations also leads to the assumption that the mobility of highly qualified persons such as scientists might be studied more fully in terms of circulatory models (see for example the investigations carried out by Kaplan, 1998; Kaplan et al., 1999; Mahroum, 1999; Meyer, 1996; Meyer and Brown, 1999). Studying the high return rates of Taiwanese and South Korean scientists who worked in the USA, Johnson and Regets (1998) coined the term brain circulation in an attempt to describe the mobility patterns found in this group. Most of these individuals returned after their stay in the USA and afterwards travelled back and forth for working reasons. The authors of the study thus concluded that such circulatory mobility enhances the exchange and cooperation between scientists across international borders. Based on these considerations and further empirical research, other scholars today assume that 'flows may not be unidirectional, homogenous, or permanent and may be mitigated by certain 'compensatory' mechanisms' (see Ackers, 2005, 99). Yet, the existing research on scientists' circulatory mobility rarely stresses gender issues (for exceptions see Ackers, 2004; Gill, 2005; Kofman, 1999; Stalford, 2005). Except for these studies, authors generally look at the distribution of mobile male and female researchers if addressing gender at all.[2]

We conclude that both venues of research potentially contributing to the issue of scientists' gendered mobility abroad – feminist migration research and research on mobile scientists – are silent about the structuring momentum that gender has in this respect. Thus, we know very little about gender differences as a determinant in issues of international mobility, its gender-specific impact on everyday life, or the respective career perspectives evolving from such experiences abroad. This chapter aims to shed light on these questions in order to appreciate, firstly, the ample variety of female migrant experiences and, secondly, the new social inequalities arising along the lines of gender, even for such economically privileged groups as female scientists in developed countries.

---

2 As Iredale (1999) points out, research is very scarce even in this respect, as data are missing even for such basic questions on gender distribution.

As I will argue, we need to adopt a research methodology that is open to the everyday life realities of the individuals under study in order to adequately reflect these gendered inequalities. In theoretical terms, an approach is required that gives space to perceive the individual as embedded in social relations such as the workplace, the family and friendship circles, as well as a broader scientific community in the case of scientists. Spatially, we cannot assume that all relevant interaction partners are located in the same physical area as our interview partners.

The empirical study

The case described in detail below is drawn from an empirical study carried out in 2002. In total, 21 scientists from Austria were interviewed who all found themselves in an early career position[3] when they went to the USA; most of the interviewees were supported by an Austrian research grant. In order to look at gender differences, the sample was selected accordingly. Also, different disciplines have been included in the empirical research with the goal of exploring the influence of differing circumstances in research. Therefore, researchers from the life sciences as well as the social sciences were interviewed in the course of the project. The data collection was done in the USA and employed the problem-centred interview method (Witzel, 1982; 1996; Scheibelhofer, 2004), in which an open, initial interview sequence is followed by a semi-structured topic guide that is flexibly used according to the interview situation (for a detailed description of this method, see Scheibelhofer, forthcoming). Towards the end of the interview, the informants were also asked to draw their most important interaction partners on a sheet of paper in the middle of which a circle indicated the position of the interviewees themselves.[4] Two different colours were used to indicate the scientists' private and professional relations. These drawings have been adopted as an additional method of data collection, since the first test interviews showed that pure oral questioning on the subject of important social relations led to poor results. Either many people were named without describing the quality of the relationship, or the focus was on a few people such that the broader picture of a social network might have been lost if these data were collected through qualitative interviewing only. Thus, the complex social relations of the interviewees with their most important interaction partners were studied based on the drawings. Not only private and work relations came to the fore; the interviewees also described different actors' spatial placements and interrelations after they had finished their drawings. The analyses of interviews and drawings were based on the Grounded Theory research strategy (Charmaz, 2000; Strauss, 1994), with a coding process leading to case studies of each interviewed person containing a chronological biography. This biography was broken down into different phases that were described

3 As Bazeley (2003) defines the concept of early career in research, the project focused on individuals who held a PhD (or doctorate in Austria) and were involved in research without holding a tenure-track position. Unlike Bazeley's early career status definition, the project did not focus on researchers who received their PhD only five years ago.

4 For a discussion of these drawings, see Scheibelhofer (2006); for a detailed description of the methods used, see Scheibelhofer (2003).

in detail according to the specific circumstances of this passage. Consideration was also given to the interviewees' reconstructed aspirations over time, the activities and realisations, and the evaluations as conveyed in the interviews (also see Witzel, 1996 for this biographic model and Scheibelhofer, 2004).

In the next step, these biographies were compared to one another and it turned out that they differed in the following dimensions: the individuals' identity constructions, social networks, professional situations, plans for mobility, and living situations. The interviews also showed that their initial plans to return after a one- or two-year stay changed as career and working opportunities subsequently took on new shapes for many of the interviewees. Mobility aspirations for the future were closely linked to the scientists' social networks. It became clear that individual social networks and future mobility aspirations were closely interlinked.

Grouping the cases under the dimensions named above, the empirical investigation identified three main forms of mobility: individuals on short-term research stays who were determined to return after a well-defined period abroad; those settling down in the USA and thus qualifying as classic work migrants; and individuals with transnational life styles based on continuous physical and communicative mobility between Austria and the USA. In the next step of the analysis, one extreme case for each type was chosen for a detailed analysis. In the following, the results of this analysis for the transnational mobility type is presented in order to make patterns of gendered transnational mobility of highly qualified women clearer for future debates.

Two centres of vital interest, two fiscal domiciles and two tax consultants: The case study of Vera Jungwirth

Vera Jungwirth[5] is a 36-year-old political scientist who had been living in New York City for ten years when the interview was carried out. After completing her MA in Austria, she obtained an Austrian grant to study at a university in New York. Two years later she received her diploma and her New York MA advisor encouraged her to apply for a PhD fellowship at that university. She won this fellowship, began her PhD studies and went into instructing undergraduates. At the same time, she was part of an Austrian-based research team carrying out independent research projects for Austrian funding institutions. Vera Jungwirth had known most of the members of this research group since her student days in Vienna. Additionally, she co-operated with two scientist friends who – like herself – lived partly in Austria, partly in New York City. At the time of the interview, Mrs Jungwirth taught at two universities in New York and at one university in Austria. Additionally, shortly before the interview took place, she won a prize for her research in Austria, thus stabilising her financial situation for another two years. After this period, she planned to allocate research funds in the USA, whilst up to that time her research funds mainly came from Austrian or European funding sources.

5 All names, dates and specifics of the fieldwork presented here have been changed in order to make the data anonymous.

The close and diverse connections with individuals both on a professional and a personal level were specific for this case when contrasting Vera Jungwirth's social relations with those of other interviewees. As her most important interaction partners, she described her husband, with whom she lives in New York City, and her family living in a small province town in Austria. After graduating from a high school in this town, Mrs Jungwirth moved to Vienna, which is one day's travel from her parents' home, in order to study. She has a very close friend, a woman from Austria, who is also a political scientist living in New York and in Austria. They often cooperate in research projects or coach one another when involved in research teams with others. Vera Jungwirth explained that another two female researchers who also move back and forth between Austria and the USA are very close friends of hers as well. Thus, Mrs. Jungwirth is part of a social network in which many participants are accustomed to moving repeatedly between two places. She pointed out that many other people she meets are also constantly on the move between their places of origin and New York, these contacts being convenient as she can consult with them about various difficulties arising from their common transnational life style. She pointed out that organisational matters are tremendous once you have employment contracts in two countries. As she has annual teaching assignments in Austria and spends part of the summer holidays there, Vera Jungwirth still rents her Viennese apartment that she shares with a roommate.

In the course of the interview, Mrs Jungwirth reflected upon her own position and saw herself as a translator between the American and German-speaking scientific communities. She defined her contribution in research as transmitting research agendas from one side to the other. At the same time, she reflected on her own agency as based on the understanding that she herself variously constructs the places between which she moves to and fro. Her activities and orientations are described as highly self-determined. Her work and social networking have resulted in an embeddedness in social relations with important scholars in her research field. While discussing the insecurities of her present situation as a researcher, Mrs Jungwirth was at the same time forging strategies of how to cope with the exigencies of the field and the circumstances she has to work in. Making use of resources (contacts, information, teaching possibilities and research funds) both in Europe and in the USA is crucial for her engagements.

Nevertheless, her economic and working situation at the time of the interview suggested uncertainty about her future plans. It was clear to her that she would not gain a tenure-track position in New York without prior research stays in other places in the USA. At any rate, she had decided that she would live nowhere else than New York City, or another intellectually and culturally important – highly competitive – city in the USA. For private reasons, going back to work in Austria was not an option as her husband, an American, did not speak German and thus would be unlikely to find adequate employment in Austria. One option the couple had discussed before from a long-term perspective was to go to London or perhaps Berlin so that both would have a chance to find professional positions in their respective fields.

Another aspect weakening Vera Jungwirth's status was that, in view of future aspirations and mobility, she was aware that she would immediately move back home to the Austrian periphery should her aged parents ever need continual care and help. As a single child, she made it clear in the interview that there was no other solution to this problem that she saw coming up sooner or later. She mentioned that in general, she avoids thinking about these insecurities, as she could not figure out how her professional and private life, with her husband being in the USA, might develop after leaving for Austria when her parents needed her. During other interviews, such gender-specific norms and views about caring for one's parents or little children proved typical for female scientists. While the issue of children and their education was also brought up by male scientists who were fathers, caring for the elderly was of no bearing to them in our interviews.

Comparing the migration experience of highly skilled with those of migrant household workers

In many respects, the migratory experiences of female scientists abroad and amply studied female labour migrants doing care work could not differ more than they actually do.

Migrant domestic workers are often seen as belonging to less minority races. In racially segmented labour markets this social stigma limits their access to work and rewards, and their social and economic contributions are consequently devalued (Lee, 1996: 10).

Being employed in private homes, immigrant domestic workers are especially removed from other workers, and tend to be less informed about their rights as workers or union politics. Often they do not receive any written work contracts and are thus in a vulnerable position vis-à-vis their employers. While the scientists studied move mostly within research exchange programs or upon accepting future employers' invitations and are consequently provided with documents such as residence and work permits in their country of residence, female migrants working in elderly and child care or as cleaning personnel frequently do not have such documents, but rather are engaged on the basis of tourist visas or as undocumented over-stayers in Western European countries. Thus, most of these women are not covered by the social security system with all its trenching consequences. The socio-economic differences between these two groups are thus considerable when it comes to their respective social status, economic situations and claims towards the nation-states they stay and work in.

Inspecting other characteristics of internationally mobile scientists as well as migrant care workers also reveals astonishing similarities. For example, as shown by Trevena's study (2005), young migrants from Poland and other Eastern European countries are mostly well educated, going abroad after studying in Poland or finishing their high school degree in order to improve their future job prospects. Similarly, studies of Eastern European women working as *au pairs* in Germany suggest that their stay abroad is seen as a means to strengthen language and personal skills in order to successfully re-enter the educational or job markets in their countries of origin (see Hess 2003). Finally, I interviewed scientists who referred to the difficult labour market situation in Austrian research institutions to an extent that it seems appropriate to surmise that leaving their country of origin and education was also a strategy to cope with the lack of employment opportunities 'back home'.

Arlie Hochschild (2002) draws a direct comparison between the outward mobility of highly qualified persons and those providing care work:

> For some time now, promising and highly trained professionals have been moving from ill-equipped hospitals, impoverished schools, antiquated banks, and other beleaguered workplaces of the Third World to better opportunities and higher pay in the First World. As rich nations become richer and poor nations become poorer, this one-way flow of talent and training continuously widens the gap between the two. But in addition to this brain drain, there is now a parallel but more hidden and wrenching trend, as women who normally care for the young, the old, the sick in their own poor countries move to care for the young, the old, and the sick in rich countries, whether as maids and nannies or as day-care and nursing-home aides. It's a care drain. (Hochschild, 2002, 17)

This comparison also indicates that mobility can only be studied in a meaningful way if all relevant spaces, networks and social groups are brought into the picture: In the above-cited case of scientist Vera Jungwirth, her parents' state of health decides whether and when she might return to Austria and probably put a hold on, or even terminate, her work as a transnationally engaged scientist. In this respect, she very much resembles migrant women shuttling back and forth between their families in their native countries and the countries where they found work: If family members are in need of care, they would readily find other arrangements that would put their achievements in forging their own transnational life styles at risk. It becomes clear at this point that women's migrational avenues – be they economically well situated or not – are still shaped by gender.

Conclusions

In view of the relevant literature in different fields of study, it has become clear that the issue of how mobility is gendered is rarely treated with respect to highly qualified people from Western Europe who move across borders. This also holds true for female scientists working abroad. Neither does the feminist literature on migration pay attention to this group – the main body of research in this area is focused on female migrants from developing countries working in the care sector – nor does the literature on researchers' international mobility address the impact gender has on individuals' decisions to leave and the further development of international careers. Yet, social scientists are called to provide insights into these questions so that social inequalities reproduced through gender hierarchies become visible to policymakers, the relevant institutions, and the broader public.

The empirical project referred to in this chapter has also demonstrated that the borderlines between short-term mobility, circular mobility, transnational life styles and migration are fluctuating – and often shift in the course of a given biography. If intervening circumstances (such as offers of employment or finding a partner of a different nationality) occur, the choices of mobility might be viewed in a different light and thus lead to changing mobility perspectives. In Mrs Jungwirth's case, the opportunity to teach and do research in more than one place – together with her close, private social relations with her family in Austria, her social circles and her husband in the USA – have guided her to forge a transnational mode of living that involves diverse insecurities but also a high degree of choice.

In order to perceive and capture such creative and fragile social realities, methodologies need to be adapted that allow a certain openness vis-à-vis the research field. As discussed above, the so-called brain drain debate is mostly based on hypothesis-testing empirical studies. As research has long failed to consider transnational work and living modes, these issues have not yet been thoroughly discussed. The reconstruction of individuals' self-perceptions and orientations have thus to be embedded into social-scientific analysis in order to explore forms of social organisation that have not yet been recognised in the scholarly discussion.

This approach allows us to acknowledge how gender gives shape to mobility and migration in groups initially not considered when reflecting on the theme of gendered mobilities. Clearly, young researchers who decide to leave Austria in order to work in the USA are in a structurally more advantageous position than many other migrants – in view of North-South and East-West divisions, as well as resulting social inequalities. But if we compare mobility experiences between female scientists and their male colleagues, we are instantly confronted with gendered perspectives. Female scientists are far more burdened with issues of child-care, partner relations and caring for the elderly than male scientists. While the latter also have to organize their family lives while abroad, gendered views of mobility are still shared by the scientists themselves, as the example of Mrs Jungwirth shows. Physical presence to provide care work is very likely still to be expected from a woman – irrespective of high qualifications and an ongoing successful career. At this point, we also detect similarities between female migrants, even if they start out from divergent socio-economic and legal backgrounds. Thus, Helma Lutz' (2003) conclusion for female migrant care workers should likewise be revisited when studying highly skilled women's mobility – gender is a structuring attribute for the experience of migration.

References

Ackers, L. (2004), 'Managing relationships in peripatetic careers: Scientific mobility in the European Union', *Women's Studies International Forum* 27, 189–201.

Ackers, L. (2005), 'Moving People and Knowledge: Scientific Mobility in the European Union', *International Migration* 43 (5), 99–130.

Allmendinger, J. (ed.) (2003), *Entstaatlichung und Soziale Sicherheit. Verhandlungen des 31. Kongresses der Deutschen Gesellschaft für Soziologie in Leipzig 2002* (Opladen: Leske und Budrich).

Basch, L. et al. (1994), *Nations Unbound. Transnational Projects, Postcolonial Predicaments, and Deterritorialized Nation-States* (Amsterdam: Gordon and Breach Science Publishers).

Bazeley, P. (2003), 'Defining 'Early Career' in Research', *Higher Education* 45, 257–79.

Bozeman, B. and Corley, E. (2004), 'Scientists' Collaboration Strategies: Implications for Scientific and Technical Human Capital', *Research Policy* 33, 599–616.

Brotchie, J. et al. (1987), 'The Transformation to an Information Society', in J. Brotchie et al. (eds), *The Spatial Impact of Technological Change* (London: Croom Helm).

Brotchie, J. et al. (eds.) (1987), *The Spatial Impact of Technological Change* (London: Croom Helm).

Callon M. (1986), 'The Sociology of the Actor-Network: The Case of the Electric Vehicle', in Michel Callon et al. (eds), *Mapping the Dynamics of Science and Technology. Sociology of Science in the Real World* (London: The Macmillan Press Ltd).

Callon, M. et al. (eds) (1986), *Mapping the Dynamics of Science and Technology. Sociology of Science in the Real World* (London: The Macmillan Press Ltd).

Callon, M. and Latour, B. (1981), 'Unscrewing the Big Leviathan: How Actors Macro-Structure Reality and How Sociologists Help Them to Do So', in Karin Knorr-Cetina and Aaron V. Cicourel (eds) *Advances in Social Theory and Methodology. Toward an Integration of Micro- and Macro-Sociology* (Boston: Routledge).

Charmaz, Kathy (2000), 'Grounded Theory', in Norman K. Denzin and Yvonna S. Lincoln (eds) *Handbook of Qualitative Research* (Thousand Oaks: Sage).

Charum, J. et al. (eds) (1998), *International Scientific Migrations Today*, ORSTOM-COLCIENCIAS, CD-ROM, Paris-Bogota.

Denzin, N.K. and Lincoln, Y.S. (eds) (2000), *Handbook of Qualitative Research* (Thousand Oaks: Sage).

Ehrenreich, B. and Hochschild, A. (eds) (2002), *Global Woman. Nannies, Maids and Sex Workers in the New Economy* (London: Grenta Books).

ETAN, (2000), *Science Policies in the European Union: Promoting Excellence Through Mainstreaming Gender Equality* (European Commission Report).

Gibbons, M. et al. (1994), *The New Production of Knowledge. The Dynamics of Science and Research in Contemporary Societies* (New York: Sage).

Gill, B. (2005), 'Homeward Bound? The Experience of Return Mobility for Italian Scientists', *Innovation* 18 (3), 319–41.

Hess, S. (2003), 'Au-pair – Sprungbrett in den Westen?! Zu einer Migrationsstrategie osteuropäischer Frauen', in Klaus Roth (ed.), *Vom Wandergesellen zum „Green Card" -Spezialisten. Interkulturelle Aspekte der Arbeitsmigration im östlichen Europa* (Münster: Waxmann).

Hillmann, F. and Rudolph, H. (1996), *Jenseits des brain drain. Zur Mobilität westlicher Fach- und Führungskräfte nach Pole*n (Wissenschaftszentrum Berlin für Sozialforschung, Discussion Paper, FS I).

Hochschild, A.R. (2002), 'Love and Gold', in Barbara Ehrenreich and Arlie Hochschild, (eds), *Global Woman. Nannies, Maids and Sex Workers in the New Economy* (London: Grenta Books).

Hollstein, B. and Straus, F. (eds) (2006), *Qualitative Netzwerkanalyse. Konzepte, Methoden, Anwendungen* (Opladen: VS Verlag für Sozialwissenschaften).

Iredale, R. (1999), 'The Need to Import Skilled Personnel: Factors Favouring and Hindering its International Mobility', *International Migration* 37 (1), 89–123.

Kaplan, D. (1998), 'Migration of the Professional, Semi-Professional and Technical Occupations in South Africa: Past Patterns, Current Trends and Policy', in J. Charum et al. (eds), *International Scientific Migrations Today*, ORSTOM-COLCIENCIAS, CD-ROM, Paris-Bogota.

Knorr-Cetina K. and Cicourel, A.V. (eds) (1981), *Advances in Social Theory and Methodology. Toward an Integration of Micro- and Macro-Sociology* (Boston: Routledge).

Kofman, E. (1999), 'The Invisibility of Skilled Female Migrants and Gender Relations in Studies of Skilled Migration in Europe', *International Journal of Population Geography* 6 (1), 45–59.

Kofman, E. (2000), 'The Invisibility of Skilled Female Migrants and Gender Relations in Studies of Skilled Migration in Europe', *International Journal of Population Geography* 6, 45–59.

Kofman, E. (2004), 'Gendered Global Migrations: Diversity and Stratification', *International Feminist Journal of Politics* 6 (4), 643–65.

Lee, S. (1996), 'Issues in Research on Women, International Migration and Labour', in G. Battistella and A. Paganoni (eds), *Asian women in migration* (Scalbrini Migration Center, Quezon City: Philippines).

Levitt, P. et al. (2003), 'Perspectives on Transnational Migration. An Introduction', *International Migration Review* 37 (3), 565–5.

Lowell, L.B. (2003), 'The Need for Policies that Meet the Needs of All', *SciDevNet*, Policy Briefs.

Lundvall, B.-Å. and Borrás, S. (1999), *The Globalising Learning Economy; Challenges for Innovation Policy* (Report based on contributions from 7 projects under the TSER programme, Brüssel, DG XII).

Lutz, H. (2003), 'Leben in der Twilightzone. Migration, Transnationalität und Geschlecht im Privathaushalt', in Jutta Allmendinger (ed.), *Entstaatlichung und Soziale Sicherheit. Verhandlungen des 31. Kongresses der Deutschen Gesellschaft für Soziologie in Leipzig 2002* (Opladen: Leske und Budrich).

Mahroum, S. (1999), *Highly Skilled Globetrotters: The International Migration of Human Capital* (Proceedings of the OECD Workshop on Science and Technology Labour Markets, DSTI/STP/TIP(99)2/FINAL).

Meyer, J.B. (1996), 'Fuite des cerveaux: comment mobiliser les compétences expatriés, Fiches Scientifiques de IRD', fiche 27 (1998).

Meyer, J.B. and Brown, M. (1999), *Scientific Diasporas: A New Approach to the Brain Drain. Management of Social Transformations* (MOST UNESCO Network, Discussion Paper No. 41, made for the World Conference of the UNESCO in Budapest).

Polanyi, M. (1962), *Personal Knowledge: Towards a Post-critical Philosophy* (London: Routledge and Kegan Paul).

Powell, W.W. and Smith-Doerr, L. (1994), 'Networks and Economic Life', in Neil J. Smelser (ed.), *The Handbook of Economic Sociology* (Princeton/New York: Princeton University Press/ Russell Sage Foundation).

Pries, L. (2002), 'Transnationalisierung der sozialen Welt?', *Berliner Journal für Soziologie* 12 (2), 263–72.

Regets, M.C. (2003), 'Impact of Skilled Migration on Receiving Countries', *SciDevNet*, Policy Briefs.

Roth, K. (ed.) (2003), *Vom Wandergesellen zum „Green Card" -Spezialisten. Interkulturelle Aspekte der Arbeitsmigration im östlichen Europa* (Münster: Waxmann).

Scheibelhofer, E. (2003), *Mobilitätsperspektiven junger österreichischer WissenschaftlerInnen. Eine qualitative Studie zur Situation österreichischer ForscherInnen in den USA* (Projektbericht IHS, finanziert durch Jubiläumsfonds der Oesterreichischen Nationalbank).

Scheibelhofer, E. (2004), 'Das Problemzentrierte Interview – Möglichkeiten und Grenzen eines Erhebungs- und Auswertungsinstruments', *Zeitschrift Sozialwissenschaften und Berufspraxis* 27 (1), 75–90.

Scheibelhofer, E. (2005), 'Mobilitätsperspektiven junger WissenschaftlerInnen im Ausland', *SWS-Rundschau* 1, 117–39.

Scheibelhofer, E. (2006), 'Migration, Mobilität und Beziehung im Raum: Netzwerkzeichnungen von InterviewpartnerInnen als interpretative Methode', in Betina Hollstein, und Florian Straus (eds), *Qualitative Netzwerkanalyse. Konzepte, Methoden, Anwendungen* (Opladen: VS Verlag für Sozialwissenschaften).

Scheibelhofer, E. (forthcoming), 'Integrating The Narration-based Interviews with Topical Interviews: Methodological Reflections of Research Practices', in the review process at the *International Journal of Social Research Methodology, Theory and Practice.*

Schulz-Schaeffer, I. (2000), 'Akteur-Netzwerk-Theorie. Zur Koevolution von Gesellschaft, Natur und Technik', in Johannes Weyer (ed.), *Soziale Netzwerke. Konzepte und Methoden der sozialwissenschaftlichen Netzwerkforschung* (München: Oldenbourg Verlag).

Smelser, N.J. (ed.) (1994), *The Handbook of Economic Sociology* (Princeton/New York: Princeton University Press/ Russell Sage Foundation).

Stalford, H. (2005), 'Parenting, Care and Mobility in the EU. Issues Facing Migrant Scientists', *Innovation* 18 (3), 316–80.

Strauss, A. (1994), *Grundlagen qualitativer Sozialforschung. Datenanalyse und Theoriebildung in der empirischen soziologischen Forschung* (München: Wilhelm Fink).

Strobl, R. (ed.) (1996), *Wahre Geschichten? Zu Theorie und Praxis qualitativer Interviews. Beiträge zum Workshop Paraphrasieren, Kodieren, Interpretieren* (Baden-Baden: Nomos).

Trevena, P. (2005), 'Polish "Intelligentsia" in London. A Case Study Of Young Graduates Working in the Secondary Sector', (Paper presented at the Conference "Vanishing Borders? Commuting Migration and Domestic Work within Europe" held in Kraków, Poland, 20–22 May 2005).

Weyer, J. (ed.) (2000), *Soziale Netzwerke. Konzepte und Methoden der sozialwissenschaftlichen Netzwerkforschung* (München: Oldenbourg Verlag).

Wichterich, C. (1999), *The Globalized Women. Reports from a Future Inequality* (London: Zed Books).

Witzel, A. (1982), *Verfahren der qualitativen Sozialforschung. Überblick und Alternativen* (Frankfurt am Main/New York: Campus Verlag).

Witzel, A. (1996), 'Auswertung problemzentrierter Interviews: Grundlagen und Erfahrungen', in Rainer Strobl (ed.), *Wahre Geschichten? Zu Theorie und Praxis qualitativer Interviews. Beiträge zum Workshop Paraphrasieren, Kodieren, Interpretieren* (Baden-Baden: Nomos).

Internet-based references

Johnson, Jean M. and Mark C. Regets (1998), International Mobility of Scientists and Engineers to the United States – Brain Drain or Brain Circulation? Division of Science Resources Studies: Issue Brief, NSF 98-316, June 22, 1998. National Science Foundation, Directorate for Social, Behavioral and Economic Sciences [website] (updated 22 November 2006) <http://www.nsf.gov/statistics/issuebrf/sib98316.pdf>.

Kaplan, D. et al. (1999), Brain Drain: New Data, New Options [website], (updated 22 November 2006) <http://sansa.nrf.ac.za/documents/tradmon.pdf>.

Chapter 9

'I'm More Sexy Here': Erotic Subjectivities of Female Tourists in the 'Sexual Paradise' of the Costa Rican Caribbean

Susan Frohlick

Introduction

Many female tourists from North America and Europe travelling on their own, with a female friend, or as part of an educational or cultural tour group are drawn to the Caribbean coast of Costa Rica. While Costa Rica in general is popular with female tourists, being ranked the world's 'third safest destination for women travellers' (after Holland and Iceland), its southern Caribbean coast appears to hold a special allure.[1] Some are drawn to 'nature' associated with the romanticised Caribbean landscape and the 'laidback vibe' of Caribbean culture, in which sexually available, hypersexualised black Caribbean men are naturalised as part of the physical landscape. The Caribbean is also an especially sensual landscape where women talk about being seduced by the bodily and multi-sensory pleasures of the beach and ocean as well as the dancing and music in the reggae bars and nightclubs.

In Puerto Viejo, a small town located near the Panama border, where I conducted ethnographic research with female tourists between 2005 and 2007, women constitute a visible group of the flows of international tourists in and out of this area. Of this group, many women participate in ethnosexual tourism, for various reasons, some of which I explain below. Nagel (2003, 200) conceptualises ethnosexual tourism as a feature of global tourism in many Third World destinations, where ethnicity, race, and sexuality converge and underpin the 'ethnically exotic sexual encounter' between locals and, in the case of Puerto Viejo, predominantly white Euro-North American tourists.[2] Women are not necessarily motivated to travel to this area by the

1 See *Tico Times*, 18 August 2006, p.52. A recent study indicates that the Pacific beach town of Jacó attracts many Euro-North American female tourists interested in Costa Rican men, some of whom travel from the Caribbean side to partake in the larger tourist economy on the Pacific (see Ragsdale et al., 2006).

2 In Puerto Viejo, sexual relations between male tourists and local women (and presumably some men) take place but are not the focus of my research, nor are such relations talked about as they are not part of the social sexual contours of the town in the same way as sexual relations between tourist women and local men.

intent to have sex with local men, but once they are there, a combination of factors incite them to do so.[3]

In this chapter, I aim to disentangle some of the complex circumstances that prompt female tourists to partake in ethnosexual tourism when they travel to the Caribbean region of Costa Rica, which some women explain as due to a shift in their erotic subjectivities and sexual identity contingent upon their own emplacement in the local landscape and crossing the border from North to South. I am interested in how white women as gendered, sexed, racialised subjects forge and assert sexual agency and identity through modern touristic travel. More specifically, I explore the claim made by some Euro-North American women travellers to the Costa Rican Caribbean – 'I feel more sexy here' – by way of an ethnographic approach that recognises travel practices and subjectivities as culturally situated yet bound up with discourses, performances and embodiments of sexuality that are negotiated through mobilities, that is, through movement across time and space and as a way of becoming (Cresswell, 2006). In this way erotic subjectivities and tourist subjectivities are intertwined and are fluid. I argue that the tension between the apparent local fixedness of black male sexuality in the natural landscape and the global mobility of white femininity can in part explain the shift in erotic subjectivity expressed by some female tourists.

Mobilities and sexualities

Zoe, a blonde haired, slim, attractive, white woman from the east coast of the United States, was one such tourist who participated in ethnosexual tourism as a contingency of her emplacement in the landscape.[4] When I spoke with her in the summer of 2005, she had travelled to Puerto Viejo as a graduation gift to herself to celebrate the completion of a university degree. She knew American friends living in the area who invited her to enjoy the Caribbean sun-sea-sand with them. Shortly after her arrival, she 'found herself' (her words) sexually involved with a local man, a man she described as 'beautiful' and 'like no man' she had ever met, a mixture of 'a macho guy' (with scars from a brawl over a woman to prove it) and a 'sweet-natured' domesticated 'jungle man' who cooked rondon soup for her.[5] When

3 I use the term 'local men' as a more inclusive term than 'Costa Rican men' because there are men living in the area who are from Nicaragua, Panama, Columbia, and men with Jamaican and Cuban backgrounds who identify themselves as 'Caribbean' rather than 'Costa Rican' (or *Tico*).

4 I use pseudonyms throughout this chapter to protect the anonymity of the research participants. I have also changed some of their identifying features including, possibly, nationality and occupation, although I have retained racial markers (that is whiteness or blackness). I apologise to any research participant who might see themself misrepresented as a result of these strategies. The interviews that appear as excerpts here were conducted between December 2004 and April 2006.

5 Rondon soup is a coconut-based stew-like local dish often made with fresh fish. Women often mention rondon soup as a kind of social currency when local men make it for them.

planning her holiday, she had not expected to get involved with anyone but was attracted to this particular man because he was 'different' (that is, from other local men and from American men). Both her sense of inadvertence (in unintentionally becoming sexually involved with a local black Caribbean man) and her perception of him as a naturalised feature of the Caribbean landscape (in the jungle cooking soup from wild native plants) drew upon linked discourses about the liberation of white femininity through travel mobility and 'primitive' 'natural' black male sexuality. Her whiteness and his blackness meshed in the context of a Caribbean vacation and sensual landscape such that their mutual desire seemed 'natural,' yet played out through particular gendered, racialised, and sexualised mobilities, as well as a much wider and complicated history of contestations over white women with black men (see Ware, 1997).

Zoe's travel story began with a stop at Wal-Mart en route to the airport in Boston to buy condoms, 'just in case'. Once she arrived in Puerto Viejo, like many other tourists she rented a bicycle to tour the quiet road alongside sandy beaches that stretch out from the main town along the coast, where she stopped to listen to howler monkeys and watch sloths that hang from the jungle canopy. She enjoyed the ease with which she moved as an unnoticed white foreign woman through the small town of just a few dirt roads and one main paved road, from her guesthouse to the beaches, the restaurants, and the bars. Bound up with this freedom of physical movement, she also experienced freedom in how she dressed. After a few days in the area, she wore revealing clothes and exposed more of her sun-bronzed body, and relished the attention directed her way when she zoomed through town on her bicycle or on foot in her skimpy attire. She took a risqué side trip to Panama with her local black lover, where as an interracial couple travelling together on vacation they caused 'a few heads to turn', as Zoe explained. These various mobilities such as the ability to move freely and easily (as a white female tourist) – by air to Costa Rica, on foot and on bicycle in the Costa Rican Caribbean town, across another Central American border by bus and water taxi, and so forth – shaped her sexual subjectivity such that part of Zoe's reason for taking a black Caribbean man as her lover, when I asked her, was that she felt 'more sexy here' (naturally) while on vacation in the Costa Rican Caribbean than she did back home in Massachusetts.

With regards to why she and many other female tourists travel to Caribbean towns in Costa Rica, part of the draw is the sexual availability of Afro-Caribbean men, as in other well-documented Caribbean destinations such as Jamaica and Barbados (see Herold, Garcia and DeMoya, 2001; Kempadoo, 2004; Mullings, 2000; O'Connell Davidson and Sánchez Taylor, 2005; Phillips, 2002; Sánchez Taylor, 2001). However, the black Caribbean men who reside in these places are not necessarily an explicit nor separate object of desire in the women's travel stories so much as they are an embodied and naturalised part of the sexualised, racialised, commoditised, and sensualised landscape of the Caribbean as a tropical paradise and fantasy playground for the pursuit of an alterity and intimacy that depends upon white mobility and black 'rootedness' (also see O'Connell Davidson and Sánchez Taylor, 2005), a point that I develop in the rest of this chapter.

Vague notions of what might await her, buying condoms 'just in case,' travelling with other women who also construct themselves as liberated, cosmopolitan feminist travellers, and then, once there, the 'discovery' that black Caribbean men desire her whiteness and body, and the public performance of her touristic freedoms, are part of the multiple, micro-mobilities that led to her sexual foray into self-described 'uncharted territory'. She had slept with black men back home, but not a black man from a developing country who did not speak her first language (English), did not own a pair of shoes, and did not eat in the same restaurants as she did. Through these mobilities bound up with her white, Western, heterosexual femininity embodied in her dress, deportment and physical movements as a tourist crossing the border to a Central American tourist destination on a American passport, her sense of sexual agency intensified. In other words, she gained empowerment through touristic mobility, a popular discourse about women's travel (for example, see www.gustywomentravel.com), but, more specifically, she gained sexual empowerment.

Zoe and other female tourists' sense of feeling 'more sexy' in the Costa Rican Caribbean is not surprising given the popular notion of touristic travel within Western modernity as a quest for transformation, pleasure, and sexual indulgence, and of mobility as central to the constitution of the modern subject (Appadurai, 1996; Cresswell, 2006; Kaplan, 2001). However, given the suturing of mobility and modernity with masculinity (Kaplan, 2001), the erotic subjectivity of female tourists (in contrast to the normative male sex tourist) is puzzling to scholars and a wider social imagination. For instance, female tourists who have sexual relations with local men are represented as sexual predators (Sánchez Taylor, 2001), as brave romantics in a new world of transnational, cross-cultural intimate tourism relations (Meisch, 1995), and as 'sex pilgrims' (Belliveau, 2006). In each framing, women's sexual agency is negotiated within the masculine hetero-normativity of travel (Enloe, 2000). Through travel, women may escape constraints from their hometown and home nation that include constraints of gender and sexuality and thus claim for themselves the ideal of the universal travel subject, who is male, white, unfettered and unmarked, in a liberal, celebratory notion of female travel mobility. Kaplan (2001) has argued that cosmopolitanism, which entails the ability to move freely about the world, is a key (and problematic) means through which personhood, subjectivity and feminist liberation are achieved in the era of modernity and globalisation. But beyond the metaphorical level, gender as embodied social relations is not escapable through mobility. Pleasure travel in particular is the realm of men and masculinity (Enloe, 2000). To travel and to participate in sexual conquest in a Third World country involves a complex negotiation in which women are not entirely freed from societal norms of white femininity neither are they permitted to take up masculine attributes (see also Jacobs, 2006; Ware, 1997). Female sex tourists thus both challenge and are situated within a trenchant masculinisation of travel mobilities in spite of an increase of women travellers in recent years. Moreover, the meanings of white femininity 'move' across space with always-sexed bodies such that women's erotic subjectivities are re-forged through travel where culturally situated discourses about gender and sex present both new 'opportunities' and old constraints.

The story of another woman, Linda, a woman in her thirties from the Midwestern United States, illustrates how sexualities (as identity, actions, performances, desires) shift and change when gendered subjects/bodies move across specific borders, locales and places. Linda had travelled to Costa Rica as a volunteer with a humanitarian tour group. During her stay in Puerto Viejo she was at first oblivious to the attention local men gave her. After a few days, she began to recognise that rather than see her as unattractive because of her size, they were attracted to her body.[6] She grew to like the attention, remaining uninterested in the men yet intent on feeling sexy and transforming her own body perception. She bought clothes that showed off her full figure as she would not do back home, and like Zoe, exposed increasingly more of her body over the course of her stay. By the time she left to return to the United States, she made a vow to herself that if she had the chance to return, she would take up the opportunity to sleep with a Rasta man, an opportunity that she felt she had let slip by on her first visit.

When she returned months later she soon hooked up with a local black man named Frank who sported short cropped dreadlocks and lived what Linda referred to, drawing upon popular ideas about Rastafarian anti-modernity, as a 'roots' lifestyle. Linda desired Frank as a 'traditional' man. By that she meant that she desired him on the basis of gender difference, that is, she liked him for his powerful masculinity, which translated to 'acting like a man' or, rather, a male who dominated in the bedroom. Her ideas about her own sexuality in terms of agency and gender performances shifted when in Puerto Viejo as a tourist such that she embraced the opportunity he presented to her to become 'more of a woman' (to accept the subordinate position) to 'accept the gender binary' (her words) and, more specifically, to satiate her sexual needs through the hegemonic, hetero-normative conventions of penetrative sex. Whereas when she was in North America, extended foreplay and oral sex were central acts to her sexual repertoire, south of the border she appeared to prefer missionary sex with a male lover who refused to perform cunnilingus and disliked foreplay. There is much more complexity to unravel in both of these women's situations but the wider point I wish to stress here is how they (re)negotiated sexual subjectivities and white femininity through what I have called 'micro-mobilities' of travel as 'liberation' that depend upon certain representations of masculinity and, more specifically, black male sexuality.

(Complicating) erotic subjectivities of female tourists

> Good girls go to Heaven, bad girls go everywhere. (Cynthia Enloe, 2000, 12)

Many women suffered anguish over the eventual demise of their relationships with local men. During the course of my fieldwork (seven months in Costa Rica in total), I spoke with approximately fifty women from Canada, the United States, France, Italy, Switzerland, Germany, Austria, England, and New Zealand, and who ranged in age from twenty to fifty years old. Their relationships varied in length from one-night affairs to one marriage that lasted twenty years. Most of the relationships were

6 Kempadoo (2004, 130–33) discusses the complex relationship between Caribbean men and 'oversize' women, suggesting that while a preference for large women may be due to the notion that they are both in need of sex and strong enough to withstand the 'vigorous' sex of a Caribbean man, at the same time, Jamaican culture is presumed to 'predispose men to liking older, larger women'.

short term. Many ended badly. At the time of her interview, Zoe was upset with her boyfriend because he had started to ask her regularly for money, and had admitted to selling crack cocaine. He was also resisting her insistence on condom use. Linda was grappling with Frank's demands that she could afford the cost of his passport application (about US $100), even though as a backpacker she was on a tight budget. The men seem to have a pull over the women in spite of the sexual health risks posed by unprotected sex as well as the financial compensation expected, which many women vehemently resist and are upset by. Such skirmishes complicate notions of women's travel as sexual liberation.

The women's stories of personal transformation and sexual self-discovery can be situated within an increasingly popular (and contentious) social imaginary of 'the female sex tourist'. Such representations of the erotic subjectivities of female tourists tend to reflect the moral discourse and contentiousness over women and mobility upheld in the bumper sticker (above) quoted by Enloe (2000). One of the key troubles posed by scholars, such as Jeffreys (2003) in her article entitled ' Sex Tourism: Do Women Do It Too?' is whether women as gendered and subordinated subjects possess sexual agency enough to warrant their inclusion in the masculinised category 'sex tourist' (see also Herold, Garcia and DeMoya, 2001). A number of films, notably 'A Winter's Tan' (1987), 'Vers Le Sud' (2005) and 'Rent-a-Rasta' (2006), contribute to an image of women who engage in transactional sex with a local man while on vacation in Third World destinations in Central America, the Caribbean and Africa as sexual predators and thus as masculinised sex tourists. On the other hand, a recent book called *Romance on the Road: Traveling Women Who Love Foreign Men* (Belliveau, 2006) extols such women as 'sex pilgrims' and constructs their unfettered erotic subjectivity favourably in terms of 'global nymphomania'. Belliveau (2006, 6) asserts, 'Casual travel sex is a leading indicator of the state of feminism and real increases in female power … they exercise a power reserved through history for male military conquerors.' In this representation Belliveau reflects a notion that links travel and sexual prowess with a liberal feminism and celebratory masculinised imperial power. By engaging in travel sex, this discourse suggests, female sex tourists wrest some of this male power and embody the 'bad girls' who 'go everywhere'. These competing representations suggest that the mobilities of touristic travel imbue feminine sexuality with a particular moral register of power.

More generally, 'being a tourist is deprecated by almost everyone' (Bruner, 2004, 7). When 'the tourist' is a woman, especially a woman travelling alone or a white woman travelling with a black dreadlocked male lover across a border in Central America, the public gaze on her mobile sexed/gendered body appears to be a scornful, suspicious glare, as Zoe's story tells. The narrator of the film 'Rent-a-Rasta' is critical of the largely unchallenged normative practice of men's use of prostitutes 'in every port', in contrast to the uproar over the relatively few numbers of women who more recently have pursued sex with native locals in various places.[7] While the situation is complex, the comment reflects the recent post-modern 'feminisation of travel' (Craik, 1997) where increasing numbers of female tourists as mobile subjects and modern consumers are demanding new kinds of commoditised

7 The original script can be heard on the trailer for the film (www.rentarasta.com).

travel experiences. Female sex tourism is situated within late-capitalist formations wherein women negotiate their erotic desires and sexual subjectivities in opposition to hegemonic masculinity and (male) 'sex tourists' (whose 'insatiable' sexual desires are naturalised and normalised) and an image of highly physically mobile women who are constructed as 'bad girls' because of the associated (and threatening) freedom of sexual expression, which for white femininity is based on 'purity' (Peake, 2007; Ware, 1997).

The women I interviewed generally did not see themselves as sex tourists. Not surprisingly, most people wish to avoid being labelled 'sex tourist' because 'the sex tourist is seen as a kind of moral monster' (Günther, 1998, 71). Women involved in this project associated the term with men who paid for sex with prostitutes, or tourists whose only travel purpose was for sex. Others did not identify even as 'tourists' given the word's particular valence as a mass-mediated experience. Women who arrived as independent backpackers or as part of a educational or volunteer group rejected the term on this basis, while others did not see themselves as tourists after remaining in the area long enough to know people and to work, or to have boyfriends.

Crucial though is how women's erotic subjectivities were constructed by various residents in the community in terms of 'whores', 'sluts', and 'loose gringas' that associated foreign women with prurience and promiscuity. A German man, the owner of a local tour business, told me: 'They're all crazy for any black dick. They can't get enough.' Another male expatriate offered the explanation that tourist women were oversexed due to the hormone-laced meat that everyone eats in the United States. Costa Rican residents also spoke of female tourists in deprecating terms. Mabel, the owner of one of the many *sodas* (small restaurants) in town and a devout churchgoer, berated what she saw as amoral behaviour of foreign women.

The perception of European and North American foreign women's erotic subjectivity as excessive and libidinous circulates widely through networks of gossip, rumour, and secrecy, and is an understanding that is gained by women who stay in the area beyond a short-term holiday (see Frohlick, 2007). 'The longer you stay, the more you learn what not to tell,' one foreign woman explained. Beyond a mere platitude, her words articulate how women's erotic subjectivities shift in movement across time and place. When women first arrive as new tourists they flaunt their newfound sexual attractiveness, not only wearing skimpy clothes and zooming through town performing and enacting liberated identities and so forth, but also flaunting their sexual 'conquests' in a public way. The longer they stay the more they know about local cultural codes and the more discreet their encounters. As Alex, a woman from New Zealand who had lived in Puerto Viejo for many years and gained considerable local knowledge about gender, sex and sexuality, explained, 'If you see a woman practically having sex with a man on the dance floor in a disco, or making out in public anywhere, even the beach, you know that *she's a* [new] *tourist.*' This is the complicated terrain upon which female tourists experience and negotiate erotic subjectivity within contentious discourses about white femininity, sexuality, and mobility.

Traveling to sexual paradise: the 'nature' of black male sexuality and white femininity

I return now to the story of Linda. She travelled a second time to Puerto Viejo because she wanted to pursue the opportunity to have sex with a Rasta man. This was a possibility opened up to her through what she saw, heard, and experienced on her first visit as an educational tourist. Although she did not travel with the intent to have sex with a local person/s on her initial trip, such intent was apparent on the subsequent trip. Insofar as 'intentionality' and 'promiscuity' along with economic exchange are intertwined in the conceptualisation of sex tourism (Günther, 1998) Linda's travels do not precisely fit this framework. For instance, she did not seek out multiple partners or transactional sex.[8] Yet I use her story to suggest that some women 'become' a kind of 'situational' sex tourist (Phillips, 2002) through place-centred gendered and racialised mobilities, where sexual agency and the quest for racialised sex is achieved through physical and mental border crossings. Such quests can be situated in normative touristic travel practices wherein erotic subjectivities of both tourist and local are infused in the landscape through embodied mobilities of travel.

Linked with the desire for alterity (cultural difference embodied through naturalised raced/sexed/gendered bodies), tourism is also a search for 'the natural' (Desmond, 1999, 145), a kind of cross-border excursion into a culture of nature. The Caribbean as 'tropical nature' is mobilised through a range of tourism imaginaries and travel practices (Sheller, 2004, 17). Like many women, Linda travelled to this area as part of a tour group that provided service to the community through humanitarian-based cultural tourism. On that trip, her intention was to experience the local culture in a more intimate way than other (mass) modes of travel would enable (see Wearing, 2001). On the return trip, her desire to become closer to nature and more intimate with the local culture had grown expressly more passionate and sexual than on the first trip. She returned to 'a sexual paradise' where male blackness was perceived to be part of the 'natural' 'sedentary' and sexualised landscape to be penetrated by mobile, white, 'loose' femininity.

Linda's description of a particularly memorable experience of her brief relationship with Frank is suggestive of this suturing of male blackness with primitive/natural sexuality and white feminine sexuality with mobility: One of the first mornings she woke up in his house, they were lying together in bed when she heard chickens and roosters crowing right near her head, inside the room. This particularly vivid aural sensory experience resonated because, in her words, 'it was totally roots'. Pruitt and La Font (1995, 431) suggest that due to 'an attraction to the powerful masculinity projected by the Rastas … men who assume the Rastafarian identity have proven to be particularly popular with the female European and American tourists with a lust for the exotic.' Linda's own attraction to a Rasta man was at least partially a desire

8 This definition of sex tourism is problematic in that it is too narrow to encompass the wide range of types of sexual relationships between locals and tourists that now take place on a global scale. 'Promiscuity' is a complex term too that requires a closer examination as a set of culturally specific moral discourses.

to relate to a poor black man of mixed Costa Rican-Jamaican ethnic identity in terms of his choice to live a pure, natural lifestyle free from material possessions and to remain close to nature (see also Pruitt and La Font, 1995).

The attraction that many North American and European women feel towards Rastafarian ideology and culture, from reggae to ganja to dreads, has been discussed elsewhere (Pruitt and LaFont, 1995; Phillips, 2005). Here I draw attention to aspects of Rastafarian culture as a 'travel experience' through which an American woman re-forges and re-orients her sexual subjectivity and erotic desires. Women viewed their sexual relations with men who bore at least the superficial significations of Rastafarian identity (dreadlocks, Bob Marley iconography, fashion) in terms of an embodied 'rootedness', as is evident in Linda's story. Frank invited her to his house – a quintessentially 'Rasta' wood-framed shack with Bob Marley and Marcus Garvey wall hangings, a brightly painted lime green front door and weathered stairs and shutters – so that he could cook her a fish dinner, which is a common 'seduction tactic' used by local men. After dinner, Linda found herself needing to dash outside in the pouring rain for a pee (for lack of an indoor toilet). This for her was a moment where she fulfilled her desire to 'experience the culture, and get as close to the insider as possible' (her words). Linda's epiphanic moment can be interpreted as an accomplishment of converging touristic pursuits for both cultural intimacy and self-transformation (Wearing, 2001), where she was able to cross a number of boundaries (racial, cultural, sexual).

On her return visit as a white woman with an 'awakened' sexual interest in the landscape as a sexual paradise in which black men are 'rooted,' she had indeed found a man with many social roots in the community. He had a close network of friends and familial ties including his mother and siblings who lived next door, and through his lack of material possessions and a 'domesticated' masculinity (cooking dinner for her) he embodied what Linda described as a 'totally roots' subject position. Linda's mobility as an independent backpacking tourist and white woman from the North both stood in contrast to and enabled her to 'consume' (the fish dinner, black sexuality, and so forth) his perceived rootedness, belonging, and sedentarism.

Sexuality as erotic intimacy extends 'beyond sex acts or desires for sex acts' (Mankekar and Schein, 2004, 358). Erotics are 'enmeshed also in fantasy, everyday practices, and social relationships' (ibid.) and, I add, modes of embodiment and movement. For Linda, the sensuousness of place mediated her subjective experience and identity formation of 'sexiness'. Many women talk about the area as a place that arouses in them an awareness of an intensified sexuality and sensuality. Expressions of feeling 'more sexy *here*' are animated through a naturalisation of black men as hypersexual that surfaces in many of the women's narratives, and is a resonating tension with their own seemingly unfettered mobility and freedom.

For Bronwyn, a yoga teacher, the paradisiacal landscape that she was drawn to as a tourist was fused with a sexual paradise; sex and the sensual are entangled. She too arrived originally as a volunteer on a community-service tour and decided to remain in Puerto Viejo indefinitely. The beach, the view, and the water were a Caribbean dream instilled in her since she was a small girl growing up in Germany. In her words,

> When stuff happens here, I can go down and sit at the beach, the beautiful Caribbean sea, have palm trees around me, this place is so beautiful. It makes it easier to deal with … here how can you be upset when you look at the view? When I'm floating about in the sea, and it's so beautiful. It means so much to me, personally, this kind of environment. Since I was a kid, this whole thing … I want to be here.

In Bronwyn's narrative, local black men were bodies that were rooted in the natural landscape (the ocean, the coastal forest) by virtue of the men's livelihoods as fishermen or nature guides. Bronwyn talked about them, many whose families migrated from Jamaica to work on the railway or on banana plantations or who were turtle fishermen in the early twentieth century (see Harpelle, 2001; Frohlick, 2007), as good genetic material, their histories of physical labour and 'natural brawn' 'good for making babies', good for looking at, and good for having 'raunchy affairs' with. She kept a list of the specific men who performed oral sex, the ones who used condoms, and the names of men she wished to avoid having sex with. These words and acts are a textual/discursive means by which the Caribbean area of Costa Rica is imagined and enacted as a sexual paradise where black male sexuality is fused onto the sensual terrain of beautiful beaches, turquoise-coloured ocean, and the fecund and 'wild' 'jungle'. The physical landscape is eroticised as the local men as 'beautiful, strong, taut, healthy bodies' are inscribed upon the natural terrain as a roster (and fantasy) of sexual resources and sensual possibilities to be gazed at, desired and consumed by white foreign women positioned as highly mobile and thus 'loose' subjects for whom the Caribbean is a dreamscape as much as a reality.

Conclusion

Unfettered sexiness has long been associated with travel mobilities, especially in the context of hot, sultry climates (see Littlewood, 2002). The Caribbean as a tropical destination is widely 'imagined as a geography of tropical enticement and sexual availability' and the promise of sexual pleasure and liberation (Sheller, 2004, 17). Many of the women travellers I interviewed took the trope of tropical paradise very seriously. Through dress, consumption, and freedom of movement, they reformulated and intensified their sexual self-expression in a specific context. When women can cross borders, enact intensified sexualities and claim, 'I'm more sexy *here*,' the sexuality associated with foreign women can also greet and partially construct them when they arrive. Sexualities are global in that people as sexual beings, along with ideas and discourses about sex and sexuality, extend beyond geo-political borders through mass media, migration, and global tourism (Altman, 2001), but sexualities play out in resolutely local communities, towns, and nations and thus are also localised.[9]

---

9 The fact that Costa Rica as a nation is bifurcated by gendered tourism mobilities and erotic desires, where male tourists travel to the Pacific to find local women and female tourists travel to the Caribbean to find local men, suggests this kind of localisation, which I do not have space to address further in this chapter.

As I describe above, in Puerto Viejo both local and foreign residents ascribe notions of promiscuity to tourist women. Such stereotypes have been forged through histories of encounters with tourists acting in ways that have sometimes clashed with local social norms (Frohlick, 2007). Stories circulate about foreign women and their sexual proclivities linked with national and ethnic identity. French Canadian women are the most 'open', while 'Canadian girls are the most shy.' 'Girls from Europe are hot, really wild, they like sex.' 'American women are rude,' yet 'American women give you the most.' The images, however contradictory, coalesce to produce a generalised popular imaginary through which female tourists' experiences are mediated. As Bronwyn explained, 'the typical naïve girl who just arrived, tourist people who come here, through no fault of their own are completely clueless as to what's going on.' One layer of mediation are travel guides for Central America that suggest a woman traveller act conservatively when travelling through small communities where religion dictates gender norms. Women are cautioned about ways to deal with the sexual lascivious nature of Caribbean and Latino men, primarily through dress. Yet when women arrive in these 'tropical paradises' they find the town full of sexually available men encouraging them to 'act like gringas', that is, sexually promiscuous, loose, and free white women. They enter into, in some ways, a ready-made 'tourist slot' that meshes with their own desires and embodies the tension between global feminist discourses of free-flowing sexual expression (Kaplan, 2001) and Latin American Caribbean discourses of a 'decent woman' (who dresses modestly, does not go out at night, does not travel alone, and so forth). As with all travel practices, the women's touristic mobilities 'bring worlds into being which are always contested terrains' (Kaur and Hutnyk, 1999, 1).

Sexual paradise is thus not only a girlhood dream, a seductive multi-sensual vacation experience, a white beach and swaying palm trees, a risqué border crossing, or a 'totally roots' fantasy of sex with a dreadlocked Rasta, but a place where women forge erotic subjectivities in the context of often confusing, emotionally overwhelming, and complex cross-cultural social relations. Linda's story ends precisely on this note. While travel to Puerto Viejo and the ensuing sex with a Rasta man enabled new formations of erotic subjectivity for her (liking her body, feeling desirable and appreciated as a woman, crossing racial boundaries, experiencing sex with a manly man, and so forth), she also became mired in complications – in refusing Frank's requests for money, she wrestled with the question of whether or not he was genuinely attracted to her, and over her own boundaries of normative white femininity where women do not pay for sex. Yet, she was able to return to the United States to resume a full-time job while her lover remained 'in place' and 'rooted', unable to pay for the application for the passport he needed in order to become a mobile, modern subject himself.

In this chapter I have sketched out some of the complex circumstances that incite white Western women to partake in ethno- (hetero)sexual tourism in the Costa Rican Caribbean, focusing on the linkages between what I refer to as their micro-mobilities and the reformation of their erotic subjectivities. The multiple movements available to them as modern cosmopolitan subjects (that is university educated, multi-lingual, well travelled, and worldly in the sense of wanting to place themselves in the world and to seek alterity) crossing borders as tourists are a means by which they acquire a particular and complex sexual identity and agency that is not available to them in their home places and nations. Mobility as a 'way of being' and 'of becoming in the world' does not take place in an abstract time and space but rather 'within specific contexts of social and cultural power' where particular mobilities are naturalised and seemingly ubiquitous (Cresswell, 2006, 2–5, 22). The claim – 'I'm more sexy here' – made by Zoe and other female tourists can be seen as implicated in touristic travel practices and discourses where crossing a border on holiday in the Caribbean region of a Central American country is itself an expression of gendered sexual empowerment. As Kaplan (2001, 235) has argued, North American and European women are liberated through travel as an invested ideology that produces modern, cosmopolitan subjects where crossing borders from North to South or from West to East animates a sexual transformation through technologies of travel. Kaplan quotes the Egyptian feminist El Saadawi, 'My face always looks more beautiful in aeroplane mirrors than it does in the mirror at home or in any mirror in the country. I don't know: do my features change simply by crossing borders, or are aeroplane mirrors of better quality?' (El Saadawi, 1993, 27 in Kaplan, 2001, 235). I argue that this sense of feeling more beautiful, more desirable, and more sexually empowered is one that is produced in part from the tension between the naturalised hypermobility of white femininity and the naturalised 'rootedness' of black Caribbean hypersexual masculinity upheld in the interracial ethnosexual encounters between white Euro-North American women and black Caribbean men. Tourism mobilities that incite sexual subjectivities are complexly gendered and racialised.

Acknowledgements

This research was funded by the Social Sciences and Humanities Research Council of Canada (SSHRC). The project would not have been possible without the generous cooperation of many people in Costa Rica who gave their valuable time and knowledge, and who shared intimate details of their lives with me, for which I am very grateful. Due to assurances of anonymity, I cannot name everyone but I would like to extend my express gratitude to Maria, Susana, Shaun, Sam, Denise, and Angela, who in different ways greatly facilitated this project. I would also like to acknowledge Jessa Leinaweaver and Shaun Sellers for careful readings of an earlier version of this chapter, as well as the useful comments from the editors of this volume, Tanu Priya Uteng and Tim Cresswell. My children, Breck and Alex, warrant special mention as always for putting up with my own often 'hyper' mobilities – without you two, I would be lost.

References

Altman, D. (2001), *Global Sex* (Chicago: University of Chicago Press).

Appadurai, A. (1996), *Modernity at Large: Cultural Dimensions of Globalization* (Minneapolis: University of Minnesota Press).

Belliveau, J. (2006), *Romance on the Road: Traveling Women Who Love Foreign Men* (Baltimore: Beaumonde Press).

Bruner, E. (2004), *Cultures on Tour: Ethnographies of Travel* (Chicago: University of Chicago Press).

Craik, J. (1997, 'The Culture of Tourism', in J. Rojek and C. Urry (eds), *Touring Cultures: Transformations of Travel and Theory* (New York: Routledge).

Cresswell, T. (2006), *On the Move: Mobility in the Modern Western World* (New York/London: Routledge).

Desmond, J. (1999), *Staging Tourism: Bodies on Display From Waikiki to Sea World* (Chicago: University of Chicago Press).

El Saadawi, N. (1999), *My Travels Around the World* (New York: Heinemann).

Enloe, C. (2000), *Bananas, Beaches, and Bases: Making Feminist Sense of International Politics*, 1st edition 1990 (Berkeley: University of California Press).

Frohlick, S. (2007), 'Fluid Exchanges: The Commodification and Regulation of Intimacy in a Transnational Tourist Town in the Caribbean, Costa Rica', *City & Society* 19 (1), 139–68.

Günther, A. (1998), 'Sex Tourism Without Sex Tourists', in M. Oppermann (ed.), *Sex Tourism and Prostitution: Aspects of Leisure, Recreation, and Work* (New York: Cognizant Communications Corporation).

Harpelle, R. (2001), *The West Indians of Costa Rica: Race, Class, and the Integration of an Ethnic Minority* (Montreal: McGill-Queen's University Press).

Herold, E. et al. (2001), 'Female Tourists and Beach Boys: Romance or Sex Tourism?' *Annals of Tourism Research* 28 (4), 978–97.

Jacobs, J. (2006), 'Tourist Places and Negotiating Modernity: European Women and Romance Tourism in the Sinai', in C. Minca and T. Oakes (eds), *Travels in Paradox: Remapping Tourism* (Lanham: Rowman and Littlefield).

Jeffreys, S. (2003), 'Sex Tourism: Do Women Do It Too?', *Leisure Studies* 22, 223–38.

Kaplan, C. (2001), 'Hillary Rodham Clinton's Orient: Cosmopolitan Travel and Global Feminist Subjects', *Meridians: feminism, race, transnationalism* 2 (1), 219–40.

Kaur, R. and Hutnyk, J. (1999), 'Introduction', in R. Kaur and J. Hutnyk (eds), *Travel Worlds* (London: Zed Books).

Kempadoo, K. (2004), *Sexing the Caribbean: Gender, Race, and Sexual Labor* (New York: Routledge).

Littlewood, I. (2002), *Sultry Climates: Travel and Sex* (London: John Murray).

Mankekar, P. and Schein, L. (2004), 'Introduction: Mediated Transnationalism and Social Erotics', *The Journal of Asian Studies* 63 (2), 357–65.

Meisch, L. (1995), 'Gringas and Otavaleños: Changing Tourist Relations', *Annals of Tourism Research* 22 (2), 441–62.

Mullings, B. (2000), 'Fantasy Tours: Exploring the Global Consumption of Caribbean Sex Tourisms', in M. Gottdiener (ed.), *New Forms of Consumption: Consumer Culture and Commodification* (New York: Rowman and Littlefield).

Nagel, J. (2003), *Race, Ethnicity, and Sexuality: Intimate Intersections, Forbidden Frontiers* (New York: Oxford University Press).

O'Connell Davidson, J. and Sánchez Taylor, J. (2005), 'Travel and Taboo: Heterosexual Sex Tourism in the Caribbean', in E. Bernstein and L. Schaffner (eds), *Regulating Sex: The Politics of Intimacy and Identity* (New York: Routledge).

Peake, L. (forthcoming), 'Whiteness', in Kitchin, R. and Thrift, N. *International Encyclopaedia of Human Geography* (London: Elsevier).

Phillips, J. (2002), 'The Beach Boys of Barbados: Post-colonial Entrepreneurs', in S. Thorbek and B. Pattanark (eds), *Transnational Prostitution: Changing Global Patterns* (New York: Zed).

Phillips, J. (2005), 'Tourist-Oriented Prostitution in Barbados: The Case of the Beach Boy and the White Female Tourist' in Kamala Kempadoo (ed.) *Sun, Sex, and Gold: Tourism and Sex Work in the Caribbean*, 183–200 (Lanhan, MD: Rowman and Littlefield).

Pruitt , D. and LaFont, S. (1995), 'For Love and Money: Romance Tourism in Jamaica', *Annals of Tourism Research* 22, 422–40.

Ragsdale, K. et al. (2006), 'Where the Boys Are: Sexual Expectations and Behaviour Among Young Women on Holiday', *Culture, Health & Society* 8 (2), 85–98.

Sánchez Taylor, J. (2001), 'Dollars are a Girl's Best Friend? Female Tourists' Sexual Behavior in the Caribbean', *Sociology* 35, 749–64.

Sheller, M. (2004), 'Demobilizing and Remobilizing Caribbean Paradise', in M. Sheller and J. Urry (eds), *Tourism Mobilities: Places to Play, Places in Play* (New York: Routledge).

Ware, V. (1997), 'Purity and Danger: Race, Gender and Tales of Sex Tourism', in A. McRobbie (ed.), *Back to Reality? Social Experience and Cultural Studies* (Manchester: Manchester University Press).

Wearing, S. (2001), *Volunteer Tourism: Experience That Can Make a Difference* (Wallingford: CABI Publishing).

Chapter 10

A Spatial Exploration of the Accessibility of Low-Income Women: Chengdu, China and Chennai, India

Sumeeta Srinivasan

Introduction

Accessibility is a concept that indicates an individual's ability to conduct activities within a given environment (Weibull, 1980). Place-based methods for measuring accessibility look at the spatial separation between key locations like home or workplaces and other locations (like grocery stores or schools) where required or desired activities occur (Miller, 2005). The challenge to planners in developing countries is that place-based accessibility varies widely for different segments of the population. For example, a person living in a location with a personal vehicle will have very different place based accessibility to jobs (or to grocery stores or shopping malls) than a person in the same location without access to a personal vehicle. This suggests that strategies for meeting accessibility needs will vary by both location and by income (and gender). To measure accessibility in this detail planners need data that disaggregate travel behaviour to the individual's travel patterns. Such data are not routinely collected in most countries. This study uses travel diary data for Chengdu, China and Chennai, India in order to understand the differences in spatial patterns of accessibility of low-income women and men.

Accessibility as a goal for planning

The relationship between travel behaviour and land use characteristics has long been a subject of debate (Boarnet and Crane, 2001). Newman and Kenworthy (1999) were among the first to study this relationship in an international context. Their study of several cities suggested a strong correlation between densities and gasoline consumption. The current debate in the context of developed countries is on the role of causality between land use and travel behaviour given that land use policies are very difficult to change. In other words, is the choice of travel mode a result of where a person lives or do people choose to live in a location based on their choice of travel mode? In contrast, low-income households in a Chinese or Indian city do not have either location or travel behaviour choices. Further, as these countries continue to urbanise, policy decisions about land use and transportation are still being made.

Decisions that can be made now could avoid the causality issues that have created an impasse between land use and transportation planning in North America. In this context, it is perhaps worth examining accessibility as a goal for planning.

Litman (2003) notes that land use patterns affect mobility and accessibility through density, land use mix, non-motorised conditions and network connectivity. Density (the number of people or per unit of land area) increases the need for common destinations that are also close, and the number of people who use each mode, thereby increasing demand for walking, cycling and transit. A mix of land uses is obtained by locating different types of activities close together, such as shops and schools within or adjacent to residential neighbourhoods) and it reduces the amount of travel required to reach common activities. The existence and quality of non-motorised conditions (walking and bicycling) can have a major effect on accessibility of households without personal vehicles. Network connectivity (many roads or pedestrian facilities that connect many geographic locations with each other) allow for shorter travel routes and more route choices. Most transportation planning is heavily focused on improving mobility rather than accessibility (Cervero, 2005). However, a major element of this focus on mobility is possibly due to the fact that accessibility is difficult to measure. It is easy for a transportation planner to measure the reduction in average travel time but much harder to measure the improvement of a household's ability to reach all possible destinations.

Measuring accessibility

Recent studies have compiled a wide spectrum of measures of accessibility (Bhat et al., 2000; Miller, 2005). Accessibility measures include cumulative opportunity measures, gravity based measures, utility measures and space-time measures.

Cumulative opportunity measures summarise the total number of jobs (or other services) with predetermined travel distance or time. This measure is sensitive to the threshold used. Therefore the gravity measure of accessibility, that weighs distance in a less arbitrary fashion, was derived (Hansen, 1959). The gravity measure of accessibility is as follows:

$$A_i = \sum_j E_j f(d_{ij}) \qquad (1)$$

Where: A is the gravity based accessibility of the location i
F is an impedance function of d_{ij}
d_{ij} is a measure of distance between locations i and j

A drawback of this measure is that it is not sensitive to variations between individuals and suggests that accessibility within a location is the same. Utility theory based estimates are more individualised measures of accessibility. Utility-based accessibility measures usually define accessibility as the expected maximum utility from a random utility model of destination choice. In this measure, the choice set includes all destinations available to a given person for a given trip purpose at a given time, travelling from a given location by a given mode (Dong et al., 2002). The advantage of such a measure is that it can be defined at the individual level. However, such measures require very detailed data documenting all the choices that are available to a resident including location, available modes, distance or time to travel to them. Space-time measures add the time dimension to accessibility estimation (Kwan, 1998). Clearly, time represents a constraint to the individual in their daily travel. They are limited in how far they can travel depending on their type of employment and other socio-economic constraints as well as the structure of the transportation system. Such estimates can better evaluate travel behaviour but are also heavily dependent on data availability at the individual level (like utility based measures). In this chapter self reported measures of accessibility were used to understand accessibility at the individual level.

The accessibility measures that capture the most complexity require very detailed location based data that are not readily available in most Chinese or Indian cities. Accessibility variations in space which also incorporate income and gender are therefore best measured when disaggregated location and individual data are available.

Context

About 10 years ago both India and China were primarily agricultural economies with over 80% of the population living in rural areas. While both countries have followed a centralised planning model for their economy the implementation of their planning models has differed considerably. Both countries implemented economic reforms that continued the opening up of their economies during the decade since 1990. A recent study indicated that cities in both countries were very similar in ways that were consistent with general processes of urbanisation, rather than the country the city was located in (Lin et al., 2004). Both countries are similar in that they are rapidly urbanising and have had rapid economic growth in recent years. However, economic growth in China has been faster than India in recent years. These differences present an opportunity for Indian cities and the relatively poor regional Chinese cities like Chengdu to avoid the urbanisation related pollution problems that have affected the largest coastal Chinese cities like Beijing and Shanghai.

Context of Chengdu, China

Since the mid-1980s, China has been pursuing a deliberate urbanisation policy as a means of stimulating both rural and urban economic development and productivity (Jun, 2003). In combination with market reforms in urban areas, cities have therefore become a very attractive place to migrate to in comparison with rural areas. The previously strictly enforced 'urban residency permits' have also been relaxed, allowing for rural-urban migration. Migration policy still favours those with higher incomes and economic prospects and a large floating population is one of the consequences of urbanisation (Shen and Huang, 2003). This floating population is still registered to their hometown, rather than the city they have migrated to, and official Census statistics do not account for them.

Chengdu is the capital of the Sichuan province and is located in the southwest region of China. It has a 2300-year continuous history. The location of the city is in a basin surrounded by hilly and terraced lands. Chengdu Municipality has 9 districts, 4 cities and 7 counties under its jurisdiction. The 1999 population of 2.4 million living in the 6 Central Districts accounted for about 25% of the Municipality's population of about 10 million, which makes it the fourth most populous Chinese city. However, of this total population, about 67% have agricultural occupations indicating that the larger proportion of the designated Municipality was predominantly rural in character. The Municipality of Chengdu has a land area of 12,300 km^2 accounting for about 12% of Sichuan province's population. The urban population in 2000 was about 3.36 million restricted to an area of about 200 sq km. The PPP-adjusted GDP of Sichuan in 2001 was approximately US$2700 per person (Chengdu Statistical Bureau, 2001).

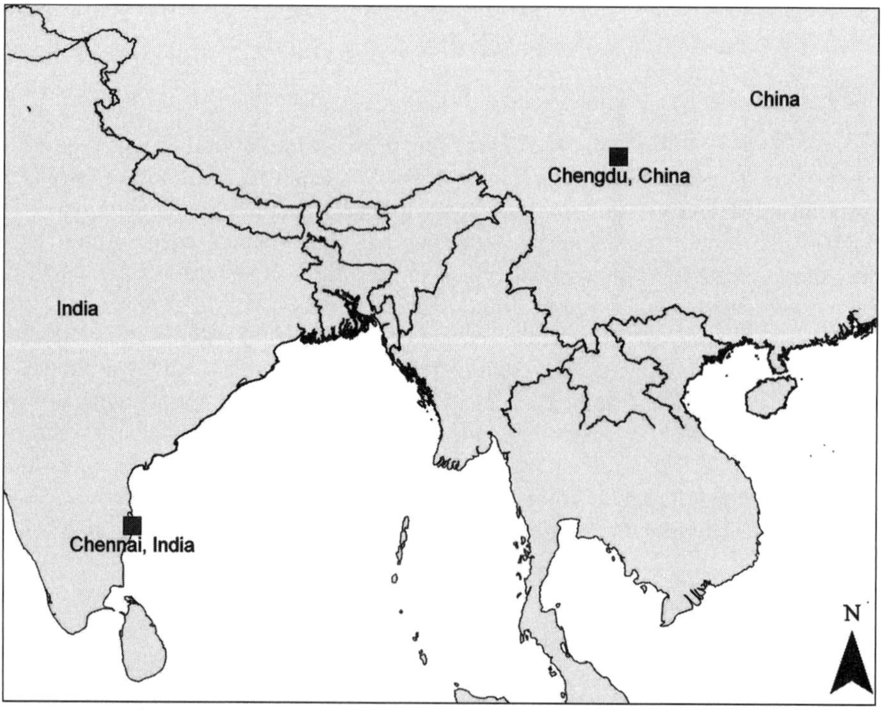

Figure 10.1 Location of Chennai, India and Chengdu, China

Context of Chennai, India

One of the distinguishing trends in urbanisation in India is that the most populous cities have grown faster than the smaller towns. In 1951 only five cities in India had populations in excess of 1 million: Kolkata (4.67 million), Mumbai (2.97 million), Delhi (1.43 million), Chennai (1.54 million), and Hyderabad (1.13 million). By 2001, however, 35 cities in India had populations of over 1 million (Census of India, 2001). Until the early 2000s the emphasis in national policy was to encourage rural development and discourage migration to cities. However, cities were the main engines of economic growth and provided better infrastructure and more job opportunities than rural areas, which encouraged rural to urban migration. A consequence of this migration has been the growth of large informal settlements in urban areas where the poor have been forced to house themselves. In 2001, the Census estimated that about 23% of the urban population of India was living in 'slums'.

Chennai is a city of 4.2 million (Census of India, 2001) in the southern state of Tamil Nadu in India. The city is spread over an area of 174 sq km and is on the Bay of Bengal on the eastern coast of peninsular India. The larger metropolitan area is spread over 1000 sq km and includes a total population estimated to be about 6 million. Modern Chennai grew out of a fishing hamlet during the early days of the East India Company. It is the fifth most populous city in India and an estimated 1 million people live in shanty-towns ('slums') in the city. The PPP-adjusted per capita GDP of Tamil Nadu was estimated to be US$1500 in 2000 (Indiastat e-Yearbook 2006).

Data

Research on the travel behaviour of those living in developing countries like India and China is relatively sparse. Most studies tend to focus on limited data collected for large-scale transportation models (CMDA, 1991; Hyodo, 2005). Furthermore, these data are restricted to one point in time (1990 for Chennai) for most Indian cities since surveys are not done at regular intervals. Most surveys do not incorporate travel diaries of detailed travel by the household members. Further the studies tend to ignore the travel behaviour of the poorer segments of the population as they are aimed at large transportation projects, which tend to serve the households with personal vehicles.

In this context, the China Project at Harvard University, in collaboration with the Research Center for Contemporary China (RCCC) at Beijing University collected travel behaviour and location characteristics data for 1001 households in Chengdu. The survey, carried out in December 2005, used a spatial sampling technique that overcomes the inability of traditional, household list-based area samples to reach rural migrants to the city (the so-called floating population). A spatial grid was created in the sample space, to create units small enough to be enumerated quickly and cheaply using GPS receivers that can identify small Primary Sampling Units (PSU) with considerable precision (Landry and Shen, 2003). Surveyors then enumerate the households residing within the boundaries of the PSU. Chengdu was divided into 769 PSU. Forty PSUs were sampled with probabilities proportional to the estimated population density of each PSU. The second ring road of Chengdu is traditionally considered as the boundary between the developed urban area and the under-developed suburban area. To include more urban residents in the sample, stratification was implemented within the second ring road. Kish Grid, which is a widely used simple procedure to select one adult from each dwelling randomly, was used. This was therefore a multi-stage, spatially stratified random sample. The data are dependent on the respondent's best estimates of their daily travel. The dataset included a daily travel activity survey of an adult in each household. This diary included 2290 trips, which recorded travel time, cost, modes used, and alternative modes available to the trip maker. Location surveys were also conducted for each of the 1001 households and survey respondents recorded travel times to regional locations like the city centre, railway stations, shopping malls, as well as local amenities like parks, markets, playgrounds, banks and post offices. The surveyor then verified the location characteristics. Although a Japanese government (JICA)-sponsored travel survey for Chengdu was carried out in 2000 (Hyodo et al., 2005), those data do not account for the floating population. From the larger sample of 2290 trips, 671 trips made by the lowest income categories reported were selected to create a sample comparable to the Chennai dataset. These trips originate from 36 PSU.

The data for Chennai come from a 2003 survey of 116 households (with a total of 509 persons conducting 1862 trips) selected through geographically stratified sampling. The 41 geographical locations are based on a 1984 study conducted by the Slum Clearance Board (a public agency in Chennai) supplemented by the author's on-site observations as the most recent Census available at the disaggregated level (1991) excludes the enormous growth in the peripheral zones of the city. There are ten zones in Chennai as designated by the local planning authority (CMDA) and are aggregated from 155 'wards' that are the most disaggregated census unit. The sampling stratified the city into 10 zones and the locations were sampled based on the relative frequency of slums in these zones. Therefore northern and southern peripheral zones, which had more slums, have more data points than other more central zones that have fewer slums. The survey recorded one working day of the week for each household and included both work and non-work activities conducted by the households. It also included location characteristics such as time taken by the households to the nearest school, grocery stores, post offices, hospitals and other amenities.

Measuring accessibility in Chengdu and Chennai

Before looking at accessibility of low-income households in the cities it is instructive to look at the similarities and differences in travel behaviour of low-income households in these cities. As Table 10.1 indicates, women in both cities make a higher percentage of trips by foot than men. Also, women tend to have a lower proportion of trips involving personal vehicles such as bicycles or other motorised modes. Interestingly, low-income women appear to make a similar proportion of their trips by transit (about 10%). The difference in bicycle use in Chengdu versus Chennai is striking but not surprising. The facilities for bicycling in Chennai are minimal. Table 10.2 suggests that the average travel time and cost characteristics of women in both cities are very similar, though men in Chennai appear to have lower average costs but spend more time travelling than men in Chengdu. The difference in the number of trips is perhaps an artefact of the differences in the way the travel diary was collected in the cities. The survey in Chennai was smaller and hence it was possible to conduct more than one interview to elicit more detailed responses about daily trips made by the household.

Other studies of household travel in Chengdu and Chennai (Srinivasan, 2004; Srinivasan and Rogers, 2005; Srinivasan, forthcoming) using these data have suggested that women in central locations with better accessibility were more likely to both make more trips and travel farther for work trips. In the case of Chennai, living in a central zone allowed for more parity in the travel costs and times of men and women. A study in West Africa by Glick (1999) also notes that the cost of transportation to the city commercial centre was key in affecting women's entry into self-employment. Travel behaviour of residents who are otherwise very similar (in terms of socio-economic status) is likely to be different if they live in locations with differing employment and transportation opportunities. Furthermore, the accessibility of a location from the perspective of a woman is likely to be different from that of a man in the same household because our data suggests that women are less likely to own a vehicle or have a license to drive it (Table 10.2).

Table 10.1 Mode choice by gender for trips in the Chengdu and Chennai sample

	Chengdu		Chennai	
	Male	Female	Male	Female
Walk	39%	59%	63%	83%
Bicycle	32%	19%	8%	1%
Transit	10%	12%	22%	14%
Personal vehicle	19%	10%	7%	2%

Note: In Chengdu males make 35% of walk trips; 57% of bicycle trips; 41% of the transit trips; 63% of the personal vehicle trips; and 45% of all trips. In Chennai males make 41% of walk trips; 83% of bicycle trips; 59% of transit trips; 80% of personal vehicle trips; and 48% of all trips.

Table 10.2 Travel characteristics by gender for trips in the Chengdu and Chennai sample

	Chengdu		Chennai	
	Male	Female	Male	Female
Average travel cost (PPP US$)	0.29	0.19	0.36	0.19
Average total travel time (minutes)	21	19	26	22
Average number of trips	2.7	2.6	4.3	5.4

Both the Chengdu and the Chennai surveys included questions about travel time to various amenities, the job and shopping centres in the city and attitudes regarding various transportation and pollution related issues. These questions were reduced to a smaller number of dimensions using principle components analysis and are further described in detail in the next section.

Accessibility and attitudinal characteristics in Chengdu and Chennai

The questions on travel time to various locations within Chengdu such as the Central Business District (CBD) of Tian Fu Square, the north and south railway stations, the closest grocery store, bank, post office, shopping malls, etc were reduced to two components. The first component was, a 'local access factor' which explained 44% of the variance, and included travel time to the nearest bank, post office and police station. The second component, which appeared to indicate regional access, explained 34% of the variance. It included travel time to the CBD and North Railway Station. Figure 10.2 indicates that the local and regional access as reported by gender varies by location. Darker shading indicates poorer accessibility in terms of time taken to travel and, in many locations, the accessibility (both local and regional access) as reported by women was darker (worse) than that of men. A principle component analysis of the attitude questions in the Chengdu survey resulted in three components. The first component, which explained 21% of the variance, included concern about accidents, rash drivers, fumes due to vehicles, and safety issues. This was noted to be a 'pro-personal vehicle' attitude. The second factor, which explained 17% of the variance, included concern about overcrowding in buses, the frequency of buses, congestion on the streets and fumes due to vehicles. This appeared to indicate a 'pro-bus' attitudinal factor. The last factor, which explained 14% of the variance, included variables that there were too many people walking on the streets and that the sidewalks were not good which can be considered a 'pro-sidewalk' factor.

The principle components of the accessibility related variables for the locations in the Chennai sample were a regional access measure (explaining 28% of the variance), which included time taken by transit to Parrys, Luz and Anna Salai, which are three employment hubs in Chennai. The next component, which explained 27% of the variance, was a local access measure which included walking time to the local market, the post office and the primary school. The third component was a health access component, which included the travel time by bus to the nearest hospital and the walk time to the nearest doctor. The attitudinal characteristics included concern about sidewalks, potholes on the roads, rash driving, speed concerns, and the price of gasoline, which appeared to indicate a 'pro-personal vehicle' attitude. The next component (pro-bus attitude) was concern about the cost of buses, overcrowding on the buses, and vehicle related pollution. The third component appears to indicate a 'pro-safety' attitude and included concern about accidents, too many people walking on the roads and traffic congestion. Overall these components explained about 73% of the variance in the sample attitudinal variables. As Figure 10.2 indicates the spatial variation in accessibility related components appears to differ by gender. Certainly some locations, especially in the northern parts of the city have poorer regional access than more central locations. Also, the southern and western periphery of the city appears to provide better overall access both in terms of regional as well local amenities than the northern peripheries.

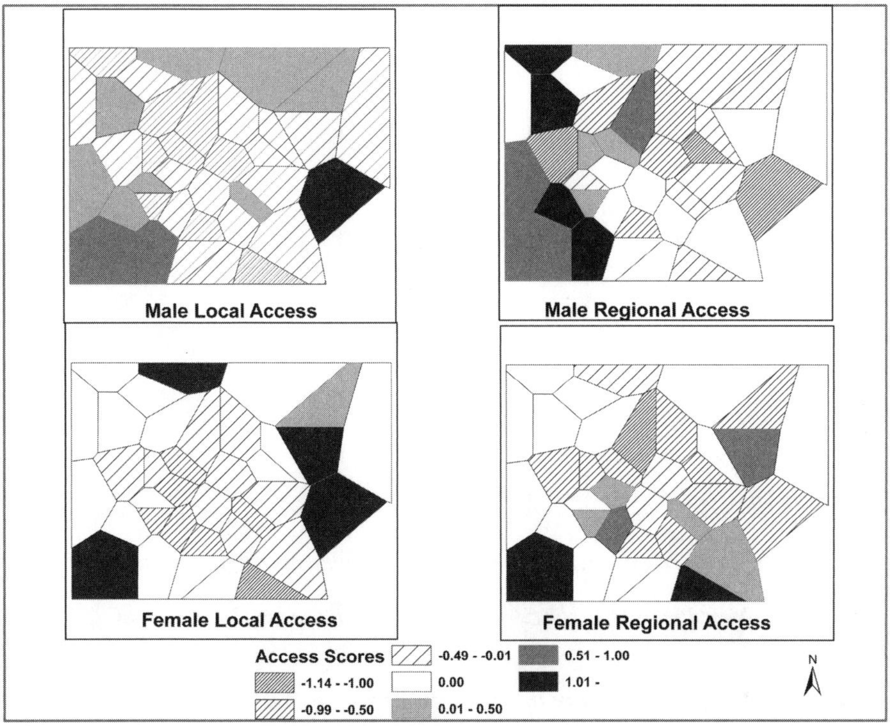

Figure 10.2 Average of regional and local access components

It is interesting to observe from the spatial distributions and the components that were extracted that there were many similarities between the Chengdu and Chennai samples. Both surveys resulted in the extraction of a similar set of accessibility and attitudinal components that varied by gender (and by income). These indicators of place-based accessibility and attitudes are at the individual level and hence indicate the differences in income and gender. As the maps in Figures 10.2 and 10.3 indicate, there are spatial differences in the regional and local access of those who live in central locations, versus those who live on the periphery. Also women and men in the same location report different levels of accessibility. In the next section the choice of mode is estimated through a discrete choice model and the accessibility and attitudinal components are tested for significance in predicting the choice of mode.

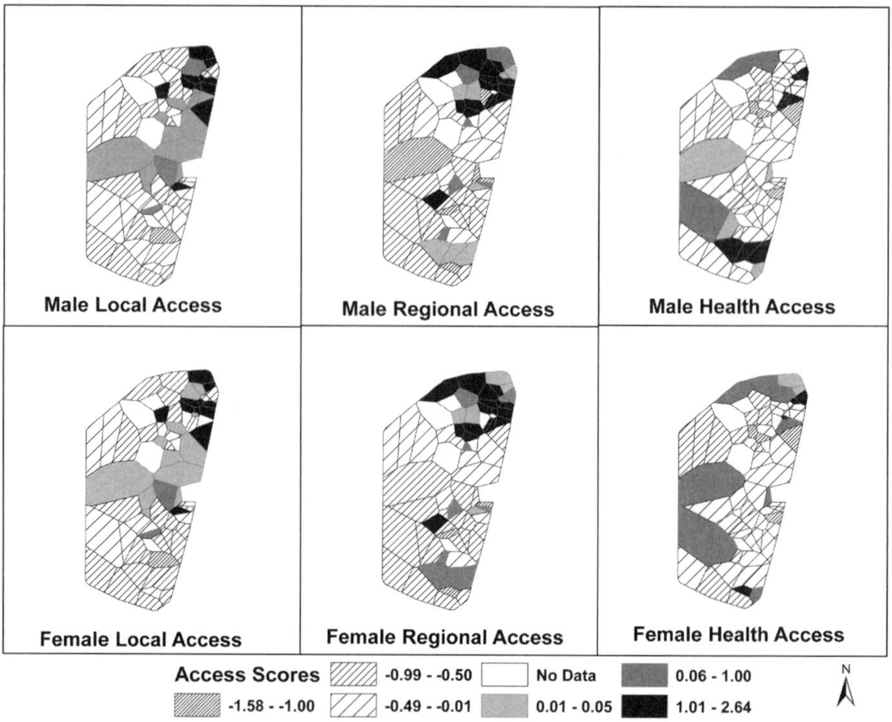

Figure 10.3 Average access scores by location and gender in Chennai

Mode choice and accessibility in Chengdu and Chennai

Utility-based indicators of accessibility have their roots in travel demand modeling. Ben-Akiva and Lerman (1979) note that, 'accessibility logically depends on the group of alternatives being evaluated and the individual traveler for whom accessibility is being measured.' They also suggest that, 'for any single decision, the individual will select the alternative which maximizes his/her utility.' Thus we use mode choice to assess the utility of using non-motorised versus motorised modes with gender, and accessibility indicators to assess if these (independent) variables play a role in mode choice.

The results of a discrete binary mode choice between non-motorised and motorised vehicles (bus or personal vehicle) are shown in Table 10.3. The results are more difficult to interpret than a linear regression model. However, the significance of gender (in the case of Chengdu women who work) and at least one measure of access in both models suggest some interesting trends. First, women do make choices that are significantly different from men's. Further, in Chennai the local accessibility of a location appears to affect the decision to use a motorised mode (worse local access means that the trip is more likely to be walk/bike). On the other hand, in Chengdu better regional access drives the choice to choose a motorised mode (worse regional access means that the trip is by foot/bike). This suggests that certain segments of the city that have poor connectivity to the rest of the city have fewer choices for employment and that these locations are especially inequitable for those who do not have access to the more expensive personal or transit modes for their trips.

Table 10.3 Binary mode choice of motorised vehicles (bus or personal vehicle) in the Chengdu and Chennai trips sample

	Chengdu		Chennai	
	Coefficient Estimate	Significance	Coefficient Estimate	Significance
Time taken on trip (minutes)	0.03	0.00	0.06	0.00
Total cost of trip (local currency)	40	0.99	0.50	0.00
Dummy indicating if trip was made by Female	NA		-0.6	0.00
Dummy indicating if trip was made by person with a job			0.4	0.09
Dummy indicating if trip was made by Female with a job	-1.56	0.00	NA	
Local Access Component	0.1	0.4	-0.3	0.01
Regional Access Component	-0.6	0.00	-0.09	0.4
Constant	-2.4	0.00	-4.5	0.00
Adjusted R Square		0.58		0.72

A spatial distribution of the logsum is shown in Figures 10.4 and 10.5 for Chengdu and Chennai. The logsum is evaluated as the natural logarithm of the exponentiated utility for non-motorised vehicle choice (Ben-Akiva and Lerman, 1979). The spatial distribution of the logsum (here darker shades indicate worse non-motorised accessibility) suggests that women have better non-motorised accessibility than men in many locations in both cities. However, the converse is also true. The map of motorised (bus or personal vehicles) would be reversed in that men tend to make this choice more often because the model predicts that it is of higher utility for them to not walk or bike in many locations.

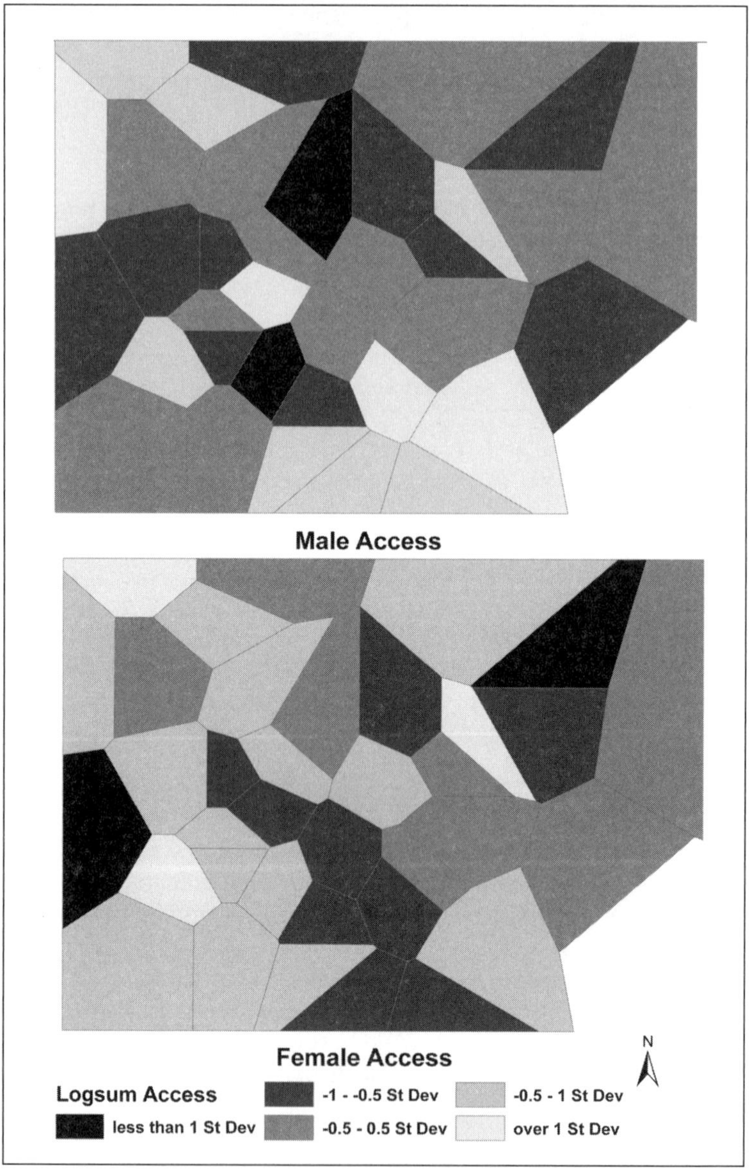

Figure 10.4 Average logsum of non-motorised mode choice by location in Chengdu

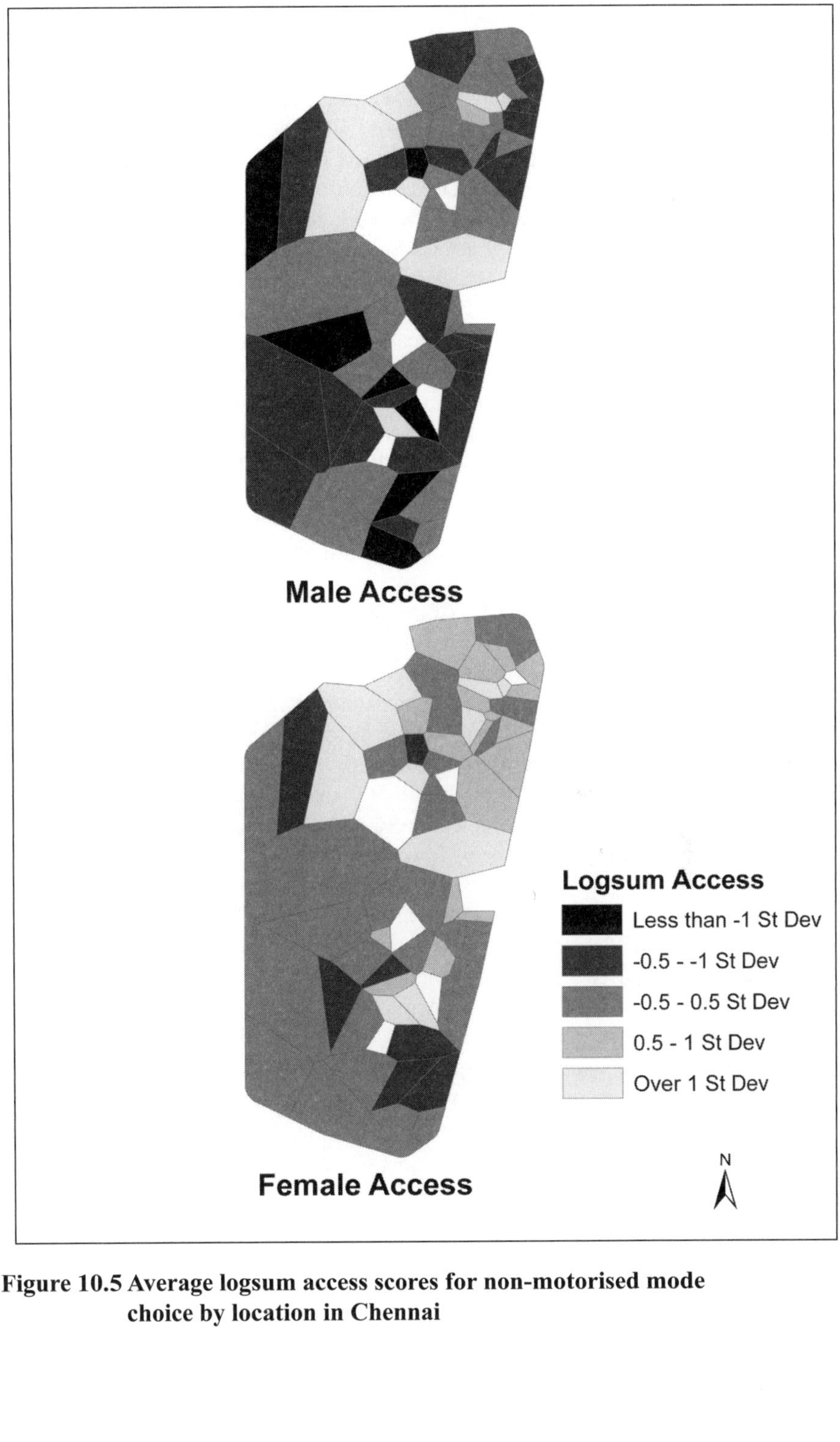

Figure 10.5 Average logsum access scores for non-motorised mode choice by location in Chennai

Policy implications for Chennai and Chengdu

Large investments in public transit like the Mass Rapid Transit System (the heavy rail under construction in Chennai) have failed to attract trips by low-income households at the levels that were predicted (Mitric, 2005). Chengdu is also currently building a heavy rail system. However, our data suggests that low-income households and, especially, women in Chennai still depend on non-motorised modes as their primary mode choice for travel to both work and non-work destinations. Further, low-income households have varying levels of access to jobs within the city. In Chennai, most jobs (even in the informal sector where low income households are more likely to find employment) are at the centre of the city or in the southern or western locations. Likewise in Chengdu, employment is located in the centre of the city or is being moved to the southern periphery (the rail lines are also being built to connect the CBD to the new location of the local government in the south of the city). These policy decisions by the local governments are likely to adversely affect those who are dependent on low cost modes such as walking or bicycles and are not able to pay for relatively higher cost transit or personal vehicles to get to work. The significance of gender and access in predicting mode choice suggests that both cities have certain access patterns in place that make it easier to continue to walk or bike depending on where the trip originates in both cities.

Most transportation planning in China and India tends to focus on mobility improvements by improving travel times through faster modes. This has resulted in expensive projects that improve vehicle speeds by building highways or rail projects. This study appears to indicate that the emphasis should be on improving accessibility, relative to the job locations (in Chengdu) and to local amenities like grocery stores, schools, etc. (in Chennai), rather than mobility. Mobility improvements tend to focus on overall reductions in travel time in the city road or transit network. In trying to improve mobility, therefore, transportation planners focus on construction of new roads or overpasses as a way of lowering overall travel times. Further, these improvements tend to favour more expensive modes like cars, two wheelers and rail transit. Low-income residents in both Chennai and Chengdu are heavily dependent on non-motorised modes or bus. Construction of roads, overpasses and heavy rail may improve overall speeds for those modes. However such construction often makes it more difficult for a local resident to cross the road to a local market, grocery store, school or a bus stop. To improve local accessibility, therefore, neighbourhood level planning will be key. This is especially true for improving the accessibility of low-income households and women, who have to walk or bike to work or for non-work activities regardless of where they live within the city.

Acknowledgments

The Chengdu survey was conducted by the Research Center for Contemporary China at Peking University and the China Project of the Harvard University Center for the Environment and School of Engineering and Applied Sciences. The Volvo Educational and Research Foundations, V. Kann Rasmussen Foundation, the Harvard

University Real Estate Academic Initiative and the Harvard Asia Center funded the Chengdu project. The American Institute of Indian Studies, Chicago, IL, funded the Chennai survey.

References

Ben-Akiva, M. and Lerman, S. (1979), *Discrete Choice Analysis: Theory and application to Travel Demand* (Cambridge, MA: MIT Press).

Bhat, C., Handy, S., Kockelman, K., Mahmassani, H., Chen, Q. and Weston, L. (2000), *Development of an urban accessibility index: literature review* (University of Texas, Austin: Center for Transportation Research).

Boarnet, M.G. and Crane. R. (2001), *Travel by Design: the Influence of Urban Form on Travel* (New York: Oxford University Press).

Chengdu Statistical Bureau. (2001). *Chengdu Statistical Yearbook 2001* (Beijing: China Statistics Press).

CMDA (1991), *Transportation Plan for Madras* (Chennai).

Dong, X., Ben-Akiva, M.E., Bowman, J.L. and Walker, J. (2002), 'Analysis of Activity-Based Accessibility', 81st Transportation Research Board Annual Meeting, Washington, DC.

Ewing, R. and Cervero, R. (2001), 'Travel and the Built Environment', *Transportation Research Record* 1780, 87–114.

Glick, P.J. (1999), 'Simultaneous Determination of Home Work and Market Work of Women in Urban West Africa.' *Oxford Bulletin of Economics and Statistics* 61(1): 57–84.

Hansen, W.G. (1959), 'How Accessibility Shapes Land-Use', *Journal of the American Planning Institute* 25, 73–6.

Hyodo, T., Montalbo, C., Fujiwara, A. and Soehodho, S. (2005), 'Urban Travel Behavior Characteristics of 13 Cities Based on Household Interview Survey Data', *Journal of the Eastern Asia Society for Transportation Studies* 6, 23–38.

Kwan, M. (1998), 'Space-time and Integral Measures of Individual Accessibility: A Comparative Analysis Using a Point Based Framework', *Geographical Analysis* 30, 191–216.

Landry, P.F. and Shen, M. (2005), 'Reaching Migrants in Survey Research: The Use of the Global Positioning System to Reduce Coverage Bias in China', *Political Analysis* 13: 1–22.

Lin, J., Srinivasan, S. and Rogers, P. (2004), *Urban Transport Infrastructure and Air Quality Characteristics: A Comparative Analysis of China and India*, Transportation Research Board Meeting, Washington DC.

Litman, T. (2003), 'Measuring Transportation Traffic, Mobility and Accessibility', *ITE Journal* (Institute of Transportation Engineers) 73, 28–32.

Miller, H.J. (2005), 'Place-based Versus People-based Accessibility' in D. Levinson and K.J. Krizek (eds.) *Access to Destinations* (London: Elsevier), 63–89.

Mitric, S. (2005), *Towards a Discussion of Support to Urban Transport Development in India.* Energy & Infrastructure Unit (South Asia Region, The World Bank).

Newman, P. and Kenworthy, J. (1999), *Sustainability and Cities: Overcoming Automobile Dependence* (Washington, DC: Island Press).

Shen J. and Huang, Y. (2003), 'The Working and Living Space of the 'Floating Population' in China', *Asia Pacific Viewpoint* 44, 51–62.

Srinivasan, S. (2004) *The Influence of Location on the Travel Behavior of Women: A Case Study in Chennai, India.* Proceedings of the Transportation Research Board Research on Womens Issues in Transportation Conference, Chicago.

Srinivasan, S. and Rogers, P. (2005) 'Travel Behavior of Low-income Residents: Studying Two Contrasting Locations in the City of Chennai, India'. *Journal of Transport Geography*, Vol. 13 (3): 265–274.

Srinivasan, S. (forthcoming) 'The Travel Behavior and Accessibility Characteristics of Women in Chengdu, China'. *World Transport Policy & Practice* (Special Issue on Gender and Mobility).

Tiwari, G. (1999), 'Planning For Non-motorized Traffic: A Prerequisite for Sustainable Transport System', *IATSS Research* 23, 70–77.

Weibull, J.W. (1980), 'On the Numerical Measurement of Accessibility', *Environment and Planning A* 12, 53–67.

Internet-based references

Census of India. (2001), www.censusindia.gov.in (last accessed December 2007).

Cervero, R. (2005), *Accessible Cities and Regions: A Framework for Sustainable Transport and Urbanism in the 21st Century*, Working Paper, UCB-ITS-VWP-2005-3 (University of California, Berkeley, USA). <www.its.berkeley.edu/publications/UCB/2005/VWP/UCB-ITS-VWP-2005-3.pdf> (Last accessed November 2006).

Indiastat e-YearBook. (2006), www.indiastat.com (Last accessed November 2006).

Jun, Tian. (2003), *Urban Slum Reports: The Case of Chengdu, China*, From Understanding Slums: Case Studies for the Global Report on Human Settlements (UN Habitat and Development Planning Unit, University College of London). <www.ucl.ac.uk/dpu-projects/Global_Report/pdfs/Chengdu.pdf> (Last accessed November 2006).

Chapter 11

Gendered Mobilities in Developing Countries: The Case of (Urban) Uganda

Nite Tanzarn

Introduction

The growing body of knowledge on transport and gender shows that there is a marked distinction between women and men's mobility, a difference that is a construct of the way society is organised and the related gendered roles, responsibilities and power (for example Tanzarn. 2006; Priyanthi and Porter, 2002; Deike; 2001; Turner and Grieco, 1998; Calvo, 1994; Bryceson and Howe, 1993). Most of these studies have provided theoretical insights into the social construction of mobility in rural areas and in developed countries. In this chapter, I explore how gender structures women and men's mobility and access in metropolitan Uganda. I also examine how transportation structures and systems create, reproduce and sustain systemic differences in material circumstances between women and men and reinforce women's exclusion and subordination. I underscore people's, especially women's, agency – acknowledging their voice, resources, awareness creation and their power to make decisions about their own lives and get involved in executing them, as a way of improving their mobility. I argue that improving women's mobility and access relative to men's has the potential to transform prevailing unequal gender relations.

Study area

The study was conducted in Kampala, the capital of Uganda and the hub of political, economic, commercial, government services and transport activities in Uganda. Located on the northern shores of Lake Victoria, Kampala's population has been experiencing rapid growth, with a population of 330,700 in 1969 growing to the current estimated population of 1.5 million. A significant proportion of Kampala's population does not reside in the city centre but in the outlying areas. This population commutes daily to and from the city centre and beyond. The chief mode of transport is by road. The city has approximately 600 km of road network, of which about 50 per cent is tarmac, the rest being gravel or earth. Most of these roads are currently in a very poor state of repair and many become almost impassable during the wet season. The city was specifically designed to expedite the movement of vehicles and not pedestrians.

The major means of transport are *matatus* (minibuses), *boda bodas* (motor cycles), private cars, special hire cars, bicycles, walking and buses. The public transport system is dominated by *matatus* carrying passengers from the suburbs. The public transport system is centred on three major parks located in the city centre. Lorries of various capacities also traverse the city roads. Motorcycles and bicycles also travel most of the city roads in increasing numbers.

Methodology

A semi-structured questionnaire was used to elicit information from a total of 225 people. These were deliberately drawn from the formal sector (23 per cent) and from areas around four major markets in Kampala (67 per cent), to represent the informal sector. Women constituted 52.4 per cent of the respondents. Wherever possible, questions were drawn from previously validated sources of transport and travel surveys (for example Department of Transport/Republic of South Africa, 2003). The questionnaire was pre-tested on the first day and necessary adjustments and refinements were made. The majority of the respondents were in the 30–40 year-old age category. Most of the women (33.1 per cent) were single. In contrast, the majority (61.7 per cent) of the men were married. As regards monthly income, 50 per cent of women earned between UGX[1] 100,000–500,000 compared to the UGX 300,000–2,000,000 earned by more than half of the men interviewed (57 per cent). In order to provide context for the more abstract data elicited through the questionnaires, I conducted in-depth interviews with six women and six men, drawn from the survey respondents, in order to deepen the understanding of why people move, where they move to and how they move.

Fieldwork data were analysed thematically according to the following questions: 1) Where do people go? 2) Why do they go there? 3) How do they go there? 4) What constrains women's and men's mobility? 5) How can these constraints be addressed?

Gendered travel patterns

The gender division of labour in urban areas is different from that in rural areas. Unlike in rural areas, urban women do not have to walk long distances in search of water or firewood. Neither do they carry heavy loads on their heads, shoulders or backs. Many of these services are provided through the market and means of transport are more widely available. Furthermore, unskilled labour is very cheap in the country (as little as US$ 10 per month) and hence almost all households in Kampala can afford hired help. Whereas urban Ugandan women and men's travel patterns do not necessarily follow the female and male gender scripts (see Tanzarn, 2006), there are nevertheless some marked differences, particularly in public spaces. This, as the discussion below illustrates, is on account of *inter alia* the differences in

1 UGX – Uganda shillings. 1 US$ is equivalent to UGX 1800.

their socially ascribed roles, their hours of work and the means of transport to which they have access.

The most frequent reason for travel for both women and men during the week is going to work. Other activities include attending school, taking children to school, shopping and looking for work. More women (30 per cent) than men (25 per cent) stay at home over the weekend. A considerably higher proportion of men (36 per cent) than women (25 per cent) work throughout the week. Activities that are mainly undertaken by women include: evening and weekend classes, shopping, and social visits to relatives, friends and the sick. Men predominate in going out of the home for recreation and taking children to school (see Table 11.1).

Table 11.1 Purpose of travel during the week/weekend by gender

During the Week					During the Weekend				
Activity	F		M		Activity	F		M	
	No. of responses	%	No. of responses	%		No. of responses	%	No. of responses	%
Work	97	82.2	91	85.0	Work	30	25.4	38	35.5
Stay home	1	0.8	1	0.9	Stay home	35	29.7	27	25.2
Evening Classes	7	5.9	1	0.9	Weekend Study	4	3.4	1	0.9
Shopping	2	1.7	2	1.9	Shopping	6	5.1	2	1.9
Look for work	0	0.0	3	2.8	Look for work	0	0.0	1	0.9
School	6	5.1	14	13.1	Work on upcountry farm	3	2.5	2	1.9
Take children to school	3	2.5	7	6.5	Visit	5	4.2	1	0.9
					Recreation/ Relax	7	5.9	14	13.1
					Pray	14	11.9	15	14.0

Empirical evidence suggests that women are largely responsible for taking children to school and to hospital as well (for example Turner and Grieco, 1998). Table 11.1 suggests otherwise. I explored some of the unexpected findings during the in-depth discussions I conducted. All the six women with whom I talked indicated that they travel with their children to school, sometimes in the company of their husbands. None of them had pointed this out during the survey, because they perceived taking children to school as one of those things they did, just like going to the market. As regards taking children to hospital, both women and men confirmed that this was a woman's responsibility, which is undertaken as and when the need arises.

As Table 11.2 shows, the responsibility for travelling with children to school, to hospital, to pray and for social visits lies with women. Men's involvement is largely limited to providing the means of transport if they are in a position to do so. While the majority of women (52 per cent) sometimes travel with their children or younger dependents, most men (54 per cent) never travel with them. Uganda's transport services do not cater to passengers travelling with children, which affects women disproportionately, restricting their ability to move freely.

Table 11.2 Responsibility for travelling with children

Do you travel with children?	F		M	
	No.	%	No.	%
Always	10	9.4	10	10.2
Sometimes	55	51.9	19	19.4
Occasionally	12	11.3	16	16.3
Never	29	27.4	53	54.1
Total responses	106		98	

Distribution of transport resources

While the majority of the people interviewed do not own any means of transport, more men than women own cars, *boda bodas* and bicycles (Table 11.3), a situation that reflects women's lack of resources. Accordingly, as Table 11.4 shows, *matatus* account for 85 and 49 per cent of women and men's means of travel, respectively. People, especially women, use a combination of means daily. Many people walk or use a *boda boda* in order to access another means, especially public transport. A *matatu* is the preferred means of transport for both women and men because it is perceived to be cheap.

Table 11.3 Ownership of means of transport by gender

Type	Female		Male	
	No.	%	No.	%
Car	7	5.9	15	14.0
Boda boda	1	0.8	6	5.6
Bicycle	6	5.1	11	10.3
None	90	76.3	75	70.1
No response	14	11.9	0	0.0
Total	118		107	

The majority of men (90 per cent) and women (61 per cent) prefer walking because it involves zero cost. The rest of the women who walk do so because they have no alternative. Walking is the only means of transport available to them either because they cannot afford transport or because they live on the margins of the city and transport services are inaccessible. As regards *boda bodas*, 52 per cent of men who use them prefer them on account of speed, while for women the key consideration for the majority (42 per cent) is affordability followed by security (36 per cent), especially at night. Whereas lower income women use *boda boda* to ensure personal security while travelling at night, relatively well-off women pay for special hire cars. In contrast, men use special hire cars because they are convenient.

Table 11.4 Most frequent means of travel

Most frequent means of travel	F		M	
	No.	%	No.	%
Walking	51	43.2	45	42.1
Bicycle	4	3.4	9	8.4
Boda boda	42	35.6	44	41.1
Private car	14	11.9	16	15.0
Matatu	100	84.7	52	48.6
Special hire car	8	6.8	7	6.5
Company transport	9	7.6	12	11.2
Total responses	220		183	

Many poor women, however, cannot afford to use *boda boda* or special hire cars. This segment of the urban poor, many of whom work in the informal sector, cope by returning home earlier than they would otherwise do. As a result, they lose out on business opportunities, as described by Sarah, a vendor in one of the biggest markets in Kampala:

> I live in Matugga, 12 km from the city. My home is about 2 km from the main road. I leave home at 5.30 a.m. every day and walk the 2 km to the stage. I deal in fresh produce: tomatoes, cabbage, eggplants and the like. I, therefore, have to be in the market by 6.00 a.m. in order to buy the produce from the traders. I have a stall in the market; however, in order to attract more customers I shift to the open market at around 4.00 p.m. Most customers don't want to shop from the inner market because it is very crowded. Business in the market is at its peak starting 5.00 p.m. This is because most of our customers do their shopping on their way back from work.
>
> Unfortunately, since I stay far away, I lose out of some of this business. It is almost impossible to get a *matatu* to Matugga after 8.00 p.m. And in any case, I cannot afford to hire a *boda boda*, which I would otherwise have to hire to take me from the stage to my home. I don't understand why but nowadays, there are many reported rape cases. This happens to especially the secondary school children who get caught up in the dark on their way from school. Even if I had the money for a *boda boda* I don't think that I would risk taking one. I am told that these *boda boda* men are potential rapists and robbers. I wouldn't mind walking if it were not so dark. But why am I complaining, even the streets of our city are pitch black at night! Walking the streets of Kampala at night is also not that safe. We hear stories, almost everyday, of muggings on the streets.

Women, especially those in the lower income bracket, spend a considerable proportion of their earnings on transport. Eighteen per cent of the female respondents reported that they spend more than 50 per cent of their income on transport, compared to only 6 per cent of the men. A higher number of men (12 per cent) than women (2.5 per cent) do not incur any transport expenses, largely because they benefit from company transport. Whereas transport is supposed to play a facilitative role in securing livelihoods, prohibitive fares intensify disadvantaged women and men's vulnerability to poverty.

Table 11.5 Proportion of income spent on transport

Proportion of Income Spent on Transport (%)	F		M	
	No.	%	No.	%
0	3	2.5	13	12.1
<10	11	9.3	14	13.1
10 –20	23	19.5	27	25.2
20 – 30	13	11.0	15	14.0
30 – 50	27	22.9	14	13.1
>50	21	17.8	6	5.6
Not applicable	8	6.8	9	8.4
No response	12	10.2	9	8.4
Total	118		107	

Interaction between gender and transport systems and structures

Gender inequalities result in the social devaluation of women who are assigned a lower position in society. Gender interacts with transport systems to reinforce inequalities between women and men. Theresa's time diary below illustrates this interaction.

Theresa's Time Diary

Yesterday, I woke up at 5:40 a.m. I had a bath and prepared breakfast for the family. Then I went to wake up my one-year-old baby just to find that she had developed a very high fever. When I got her out of bed, she started crying. She cried so hard that everyone in the house came to see what was happening. That was at about 6:40 a.m. and my husband was in the bathroom. When he got out of the bath, he started quarrelling and abusing me, as if I was the cause of the child's pain. I, of course, had to change my travel plans. I prepared the baby for hospital. We left home 30 minutes later than the usual 7.00 a.m. and we were thus caught up in a traffic jam for close to one hour. My husband hadn't let up on his ranting. He seemed to blame me for everything: the child's illness, our subsequent leaving home late and the slow moving traffic as well!

At 8:45 a.m., my husband dropped me at the hospital. He said that he couldn't come in with me because he was already late for work. I called the office to let them know that I wouldn't be reporting for work because my child was sick. After examining the baby, the doctor said that they had to admit her for close monitoring. I had to go back home to get bedding and other essentials. I called my husband but he refused to take me saying that he had already wasted too much time travelling. Since I didn't have money for a special hire car, I got onto a *boda boda*. Whereas this was risky, it was much faster than a *matatu*. I made the return trip in 45 minutes. On arrival back at the hospital, the nurse told me that the doctor had recommended a diet of fruits and milk for the baby. I got onto another *boda boda* and went to the nearby trading centre.

I got back to the hospital at around 1.00 p.m., tired, dusty and very hungry. And that is when my husband called to find out how the baby was doing! Fortunately, she had responded to the treatment and the doctor promised to discharge us by the evening. We left hospital at 5.20 p.m. and it took us close to two hours to get home. I remembered to buy our daily home supplies after we had passed the shops. My husband refused to turn back arguing that he had had a very long, tiring day! I left the baby with the house maid and walked back to the market. Since I was very tired, I decided to take a *boda boda* back home. My husband was furious saying that I was a very wasteful and lazy woman!

Constraints on women and men's mobility

Transport infrastructure and services play a facilitative role in women and men's mobility and access; however, they can also impose physical restrictions on their ability to move. Both women and men identify the following as the greatest transport-related constraints on their mobility: 1) the poor state of repair of roads; 2) traffic jams, especially at peak hours; 3) traffic accidents; and 4) high costs of transport associated with the ever-increasing prices of fuel (see Table 11.6). Other issues identified include lack of security and congestion on the road and disrespect for traffic regulations. Women also perceive vehicles being overloaded and in poor mechanical condition as an issue that influences their freedom of movement. Corruption is another factor. Women argue that the traffic police department condones traffic offences such as reckless driving and overloading. This, coupled with the awarding of maintenance and rehabilitation contracts to incompetent civil engineering contractors, contributes to the economic and social inefficiencies in the transport systems and structures. Women also complain of sexual harassment from rude and/or smutty public transport operators. Of the 100 women who use *matatus*, 93 reported that they had been touched on different parts of the body by either the conductor or the driver.

With regard to how transport means shape mobility, both women and men identified issues similar to those pertaining to transport in general. For *boda bodas,* reckless driving was identified by both women (33 per cent) and men (26 per cent) as the biggest challenge, followed by a lack of security for individuals (cyclists and their passengers) and property. There are marked gender differences in the problems associated with walking. Whereas women identify lack of pedestrian walkways and disrespect for pedestrians by other road users as the biggest challenges, a lack of security on the streets, especially at night, was prioritised by the men as the key constraint followed by the overflowing sewage and rubbish on the road (see table 11.7).

Table 11.6 General mobility constraints

Mobility Constraint	% F (n=118)	% M (n=107)
Roads in poor state of repair	42	37
Traffic jams at peak hours	39	37
Traffic accidents	33	30
High transport costs/fuel costs	31	31
Vehicles in poor mechanical condition	18	3
Lack of security on the road	14	9
Too many vehicles on the road/congestion	14	9
Overloading	14	6
Disrespect for traffic regulations in general	12	8
Reckless driving/excessive speed	12	8
Narrow roads/no parking space	9	8
Corruption	9	1
Environmental pollution	9	2
Rude/Smutty public transport operators	8	2
Unreliable transport services	6	4
Inadequate transport services	6	8
Poor planning	6	5
Travel time too long	4	2
Inadequate road signs/markings	4	5
Lack of pedestrian walkways	3	2

Table 11.7 Ranking of mobility constraints by means of transport and gender

Constraint	Vehicles		*Boda Boda*		Walking	
	F	M	F	M	F	M
Accidents	1	1				
Traffic jam	2	2				
Vehicles in poor/dangerous mechanical condition	3	3				
Roads in poor state of repair	4	3				2
High transport costs (fuel, maintenance)	5	3				
Reckless driving	6		1	1		
Lack of security on the road		6	2	2		1
Inadequate traffic regulations and enforcement			3		3	
No *boda boda* routes				3		
Disrespect for traffic regulations			4	4		
Lack of pedestrian walkways					1	4
Disrespect for pedestrians by other road users					2	5
Inadequate roadside shelters					4	
Time consuming/tiring					5	3

The ability to move around without restraint is also intrinsically linked to gender relations and the roles and responsibilities that form the basis of gender differences. In addition to the physical barriers to mobility that are related to transport, there are gender-imposed constraints that affect more women than men. The first two cases below are examples of husbands' control of their wives' mobility. The third case illustrates the double standards as regards spousal control.

Case I

I am a 45-year-old professional woman. I am what many would consider a privileged woman, an empowered woman. I am well-educated (PhD holder), married with children and I have a good job. There is one thing I have failed to do though…drive a car. And at my age, I think that it is now something I will never be able to do. My husband uses the family car (which we both contributed to) to control me. This is not so much of a problem when I am going out to work because I have access to company transport. However, when I need to go out shopping or visiting, it becomes an issue. Many times, my husband refused to take me. He sometimes insists that I use public means. The problem is that we live about 3 km away from the main road and the road is very muddy during the wet season. If the journey is very urgent, I hire a taxi. My family and friends say that since I got married, I no longer have time for them!

Case II

My movement is controlled by my husband and my boss. My husband expects me to inform him of all my movements. I have to let my husband know where I am going, even if I have to go out of the office for brief moments only. What would I say if he called and I was away from the office? He has to know where I am at all times. I don't necessarily have to ask him for permission in order to go anywhere but as a responsible wife, I have to inform him where I am going and why I am going there. I don't want him to be suspicious of me. If my husband is to tell me anything about his movements, he will do so only if I ask him on his return. But I suspect that many times he doesn't tell me the truth.

Female banker

Case III

Ask permission from whom? I don't have to seek anybody's permission in order to go anywhere. If I am in a good mood, then I may inform my wife of my whereabouts. This is especially so if I am going to be late or when I decide to spend the night away from home. Why would I need to inform my wife of anything anyway? But I expect my wives to get permission before they go out of the house apart from of course when they are going out to work. A wife just leaving home without permission? That is unheard of! That wouldn't be a dutiful wife.

Businessman with two wives

'Missed' journeys

Both women and men have journeys that they would like to make but cannot. The greatest percentage of 'missed' journeys are social in nature – many would like to visit their parents, go to bury their dead, take children to hospital or visit the sick, but they cannot. While 40 per cent of men would have loved to go for a holiday or a tour, only 29 per cent of women prioritised this as a missed trip. Whereas 7 per cent of women perceived transport as constraining their shopping, men did not see this as a missed journey, but instead saw it as limiting their business opportunities. This difference possibly arises out of the fact that women are responsible for shopping and the generally more mobile men travel away from homes for business. Women's relatively lower emphasis on holidays/leisure could also be attributed to their lack of time for leisure as well as the resources to go off touring.

Table 11.8 Missed journeys

Missed Journey	F		M	
	No.	%	No.	%
Formal work	6	5.4	4	4.0
Shopping	8	7.2	0	0.0
Business	8	7.2	16	15.8
Leisure/holiday/tour	32	28.8	40	39.6
Social	57	51.4	41	40.6
Total responses	111		101	

Women and men identify similar reasons for missed journeys. However, there are gender differences in the extent to which these constrain their mobility. The study confirmed that relative immobility is a defining feature of poor women and men. The prohibitive cost of transport appears to be the major factor limiting both women (44 per cent) and men's (61 per cent) mobility. An additional dimension of poverty for women is lack of time for travel. This is on account of their relatively high involvement in household work.

Table 11.9 Reasons for the 'missed' journeys

Reasons for missed journeys	F		M	
	No.	%	No.	%
High transport costs	52	44.1	65	60.7
Accidents	20	16.9	15	14.0
Roads in poor state of repair	1	0.8	13	12.1
Lack of time	33	28.0	4	3.7
Lack of security in some parts of the country	5	4.2	2	1.9
Others*	7	5.9	8	7.5
Total responses	118		107	

*For women, 'others' include their husband's control of their movement; and for the men: traffic jams and inadequate means of transport.

Addressing constraints on women and men's use of space over time

Poverty appears to be a defining factor of mobility for women especially. In order to improve their mobility, 39 per cent of the women interviewed believed that they should be economically empowered in order to be able to better cope with the ever-rising costs of transport. Construction of walkways in order to provide safety for pedestrians is another priority. Other suggestions for improving mobility include regulating speed and adopting inclusive and effective transport policies including regular road maintenance. Additional measures suggested by women to improve mobility included ensuring safety on the road through enforcing traffic laws related to appropriate parking and overloaded vehicles, as well as regulating drunken driving. In contrast, men are more interested in constructing new roads/tunnels/flyovers/bypasses in order to reduce travel time, and the subsidy of fuel and restrictions on the importation of used vehicles (see Table 11.10).

Table 11.10 Suggestions for addressing mobility constraints

Suggestions for improvement	Women		Men	
	No.	%	No.	%
Traffic/Road education	16	13.6	4	3.4
Construct new roads/tunnels/flyovers/bypasses	19	16.1	19	16.1
Regular and effective road maintenance	23	19.5	15	12.7
Reduce costs of fuel/subsidise fuel	16	13.6	14	11.9
Restrict importation of used vehicles & license only roadworthy vehicles	32	27.1	21	17.8
Pedestrian walkways	38	32.2	31	26.3
Regulate drunken driving and speed (governors, cameras)	56	48	34	28.9
Regulate transport fares	9	7.6	3	2.5
Enforce traffic laws (parking, overloading, talking on phone)	22	18.6	10	8.5
Effective transport/city planning/enlarge city	37	31.4	22	18.6
Promote other modes of transport/buses	17	14.4	11	9.3
Reduce poverty (credit, employment creation, income generation)	46	39.0	24	20.3

Concluding remarks

This study analysed the travel patterns and transport needs of Ugandan women and men in the greater Kampala area who travel from their homes regularly. It examined the influence of transport systems and structures on women and men's mobility and access. The study established that mobility is vital to both women and men in securing their livelihoods, enhancing their human and social capital and in fulfilling their societal obligations. The factors that structure women and men's patterns and intensity of travel include affordability, reliability, safety and frequency of transport. An additional gender-imposed barrier for women is the freedom to move.

Whereas space is not intrinsically gendered, the inequitable positioning of women in society relative to men structures women and men's use of urban space over time and therefore results in exclusion. Whereas there are no overt measures to exclude women, omission of the consideration of gendered travel patterns, and therefore transport needs, results in the exclusion of women, in missed trips and therefore missed opportunities.

Women and men identify the same physical barriers to their mobility, barriers that are associated with transport infrastructure and services. However, their prioritisation of these transport-imposed barriers, and therefore needs, is sometimes different. This has implications for transport policy and planning. If transport policies, plans and programmes are to be economically and socially efficient, if they are to benefit women equally with men, then they must be informed by an analysis of women and men's travel patterns and their related transport needs.

In practical terms, this means institutionalising the collection and analysis of sex and gender disaggregated data in routine monitoring, as well as in household, travel and transport surveys. Many aspects of gender are not easily quantified; therefore survey data should be complemented by more qualitative analyses of women and men's mobility and access.

References

Bryceson, D. and Howe, J. (1993), 'Rural Household Transport in Africa: Reducing the Burden on Women?' *World Development* 21, (11), 1715–28.

Calvo, Christine Malmberg (1994), 'Case Study on Intermediate Means of Transport Bicycles and Rural Women in Uganda', SSATP Working Paper No.12.

Deike, Peter (2001), *Gender and Transport in Less Developed Countries*. Background Paper for the Expert Workshop 'Gender Perspectives for Earth Summit 2002: Energy, Transport, Information for Decision-Making' Berlin, Germany, 10–12 January 2001, <http://www.cityshelter.org/13_mobil/23tend.htm> accessed 16 August 2006.

Department of Transport/Republic of South Africa (2003), *National Household Travel Survey 2003*, Technical Report.

Priyanthi, Fernando and Porter, Gina (eds) (2002), *Balancing the Load: Women, Gender and Transport* (London: Zed Books).

Tanzarn, Nite (2006), Gender Survey of Ongoing Transport Projects and Programs, Uganda Case Study prepared for the World Bank Sub Saharan Africa Transport Policy Program (SSATP).

Turner, Jeff and Grieco, Margaret (1998), *Gender and Time Poverty: The Neglected Social Policy Implications of Gendered Time, Transport and Travel*, paper presented at the International Conference on Time Use, University of Luneberg, Germany, April 1998.

Internet-based reference

Grieco Margaret (2005), *The walking bus: the gender implications*, <http://www.geocities.com/gender_and_transport/braunschweiginaugural.htm?2006> accessed 17 August 2006.

Chapter 12

Gender Differences in the Influences of Urban Structure on Daily Travel

Petter Næss

Introduction

In a booklet published by a group of Norwegian feminist transport researchers and planners in the 1980s (Jenseth et al., 1986), two different models for future urban development were discussed: the 'masculine city' and the 'feminine city'. The 'masculine city' was based on development where women increasingly adopted men's mobility patterns, with the road network and urban land use adapted to steadily growing automobility. The 'feminine' model implied that men, through encouragement and coercion, increasingly adopted women's traditional travelling patterns, with shorter trips and a greater number of trips on foot, by bike and on public transport. And instead of a decentralisation of dwellings, workplaces and service facilities to the outer areas of the city, this model implied urban densification, which contributed to a reduction of the travelling distances between daily destinations.

Several studies have shown that men often choose among job opportunities from throughout a metropolitan area, while women, to a greater extent, confine their choice of workplaces to those available locally (for example Jørgensen, 1992; Hanson and Pratt, 1995; Hjorthol, 1998; Lee and McDonald, 2003). Given such a gender difference, women will, to a greater extent than men, find employment in local centres, whereas men will more often orient themselves towards the main metropolitan-level concentrations of workplaces. Moreover, if the husband is regarded as the spouse with the greater right to use the car, males may be encouraged to travel by car regardless of whether or not alternative means of transport can be used without too much inconvenience, whereas women, to a greater extent than men, may become used to travelling by transit or non-motorised modes.

For urban planning, such differences are important both in order to assess the welfare impacts of alternative land use and transport policies, and in order to anticipate how changing gender roles are likely to affect future relationships between the form of the urban landscape and travel. However, most studies of relationships between urban form and travel have addressed the situation among the urban population at large, without investigating gender differences.

This chapter presents results from one of the few studies that have investigated how residential location influences daily travel differently among women and men: a comprehensive study of residents of the Copenhagen Metropolitan Area, Denmark. A more complete account of this study is given in Næss (2006a). The investigation has been informed by concepts and perspectives from time geography, location theory and mobility sociology. Gender differences in travel are partly seen as derived from differences in activity patterns and responsibilities and partly as emerging from differences in access to, as well as attitudes towards, different modes of travel. These differences are again seen as being largely the result of socially constructed gender roles.

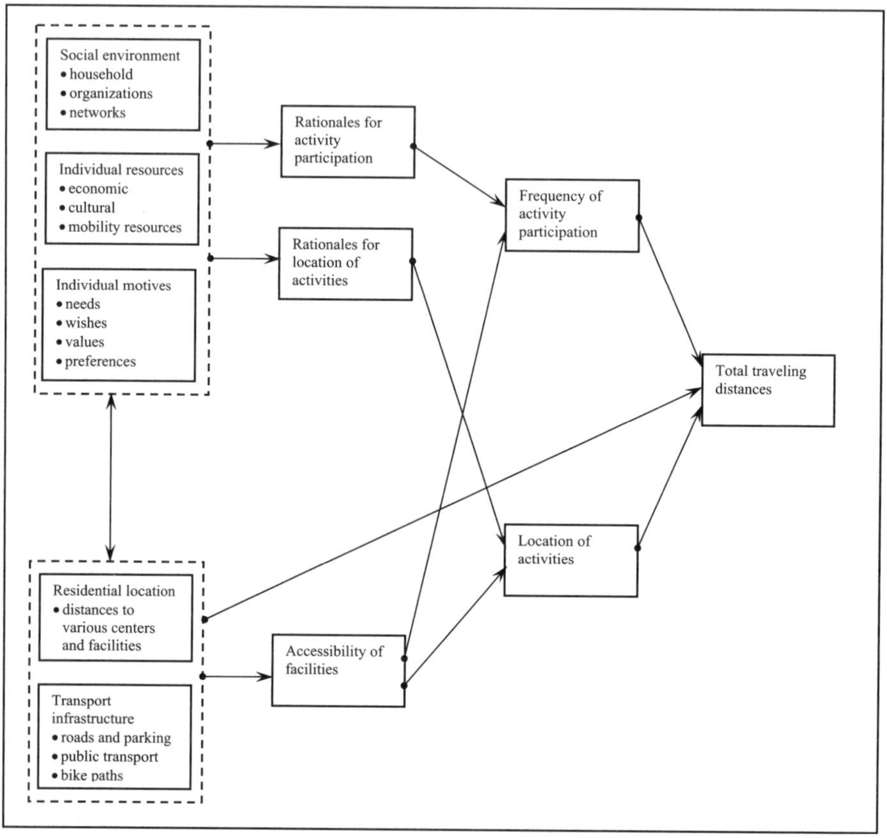

Figure 12.1 Behaviour model showing the assumed links between urban structural, individual and social conditions, accessibility to facilities, rationales for activity participation and location of activities, actual activity participation and location of activities, and total travelling distances

Figure 12.1 shows a simplified behavioural model of the ways in which individual, urban structural and other social conditions are assumed to influence daily travelling distances through accessibility to facilities, rationales for activity participation and location of activities, frequencies of activity participation and actual location of activities.[1] The location of the residence relative to various centres and facilities, combined with the transport infrastructure on the relevant segments, determines how accessible these centres and facilities are from the dwelling. Accessibility will be greater with a lower *friction of distance*, where the latter is a function of time consumption, economic expenses and inconvenience involved when travelling from one place to another. All else being equal, accessibility will of course be greatest for the closest facilities. However, what determines the most accessible location varies with travel modes, which are dependent on the layout of the public transport network, the driving conditions along the road network, and the conditions for walking and biking, among other factors.

The residents' individual resources, motives and social environments influence their rationales for participating in activities (including their trade-offs between motivation for participation and friction of distance) and location of activities (notably their balancing between proximity and choice). Combined with the accessibility of various facilities, these rationales influence the frequency of activity participation as well as the actual locations chosen for the various activities. The total distance travelled is a consequence of the geographical locations chosen for the activities in which the resident participates, the distance along the transport infrastructure network from the residence to these locations, and the frequencies at which the various activities are carried out.[2]

The chapter first briefly outlines the methods used in the Copenhagen Metropolitan Area study and then presents empirical results. The focus has been confined to weekday travel in the evaluation of the influences of residential location on travelling distances and modes. This is partly because time budgets are tighter and the competition between household members about the right to use the household's mobility resources more pronounced on weekdays, and partly because the relationships between urban structure and travel are generally clearer on weekdays than at the weekend. Gender differences in travel behaviour – and in the influences of urban structure on travel – are therefore more easily identified in weekday travel. The final part of the chapter raises the question of how urban planning can contribute to mobility patterns and mobility opportunities that counteract the barriers to women's equal participation in social life while also contributing to more environmentally sustainable urban mobility.

1 The figure does not show conditions influencing the travel modes used, which make up another important aspect of the study. Travel modes could be expected to be influenced indirectly by the factors shown in Figure 12.1 through the influence of these factors on travelling distances, and directly by individual resources and motives, transport infrastructure and social environments.

2 This presupposes that the residents choose more or less the shortest routes. Our qualitative interviews show that this is to a large extent the case for daily travel, and any deviations from the shortest route are usually quite small.

Data and methods

The Copenhagen Metropolitan Area study included a large travel survey among inhabitants of 29 residential areas, a more detailed travel diary investigation among some of the participants of the first survey, and qualitative interviews with 17 households. The chosen residential areas included a wide range of urban structural situations, varying in their location relative to downtown Copenhagen and lower order city centres, as well as in their density, composition of housing types and availability of local green areas. The participants in the qualitative interviews were recruited from three inner-city residential areas, one peripheral area close to an urban rail station and one peripheral area with very poor public transport services.

The questionnaires included questions about travel behaviour, activity participation, socio-economic characteristics of the respondents, attitudes towards transport and environmental issues, and possible obligations, social relations or routines likely to influence travel behaviour. The main survey included questions about the distances travelled by each mode on each day during a week. The travel diary investigation covered the 4-day period from Saturday morning to Tuesday night, and included detailed questions about each trip made (purpose, length, travel time and travel mode). The qualitative interviews were semi-structured, focussing on the interviewees' reasons for choosing activities and their locations, travel modes and routes, as well as the meaning attached to living in or visiting various parts of the city. The interviews, usually lasting between one-and-a-half and two hours, were tape recorded and transcribed in their entirety.

Most of the presentations of gender differences below are based on the quantitative parts of the material. The qualitative interviews were designed to shed light on the more detailed causal mechanisms by which residence location influences the interviewees' travel behaviour, but they did not focus specifically on gender issues. However, the material from the two surveys provides a base for a number of interesting comparisons between male and female respondents.

Many of the results are presented in the form of simple comparisons between men and women living within different distance belts from the Copenhagen city centre. However, more sophisticated statistical analyses have also been carried out in order to assess the separate effects of different potential influencing factors. In most of these analyses, four urban structural and 19 socioeconomic, attitudinal or other control variables were included. The four urban structural variables were: Location of the residence relative to downtown Copenhagen, relative to the closest second-order urban centre and relative to the closest urban rail station,[3] and density of inhabitants and workplaces within the local area of the residence.[4] Distances to downtown Copenhagen and the closest second-order centre reveal something about the location of the residence in relation to the metropolitan-scale hierarchy of

3 In the statistical analyses, the linear distances measured along the road network were transformed by means of non-linear mathematical functions taking into account the fact that the attraction of a centre decreases if the centre is too far away from the residence.

4 Calculated as the sum of inhabitants and workplaces per hectare within a circle with a radius of 800 m around the centroid of the residential area.

centres. The distance to the closest urban rail station indicates the accessibility of the residence to the main public transport system of the region (and to the local centres usually located around the stations), whereas local area density says something about the population base as a factor for local facilities.

Activity participation

Women's generally restricted mobility resources put constraints on their radius of action, compared to men. Living far from relevant facilities could therefore be expected to result in a higher reduction in activity participation among women than among men. However, our material from the Copenhagen Metropolitan Area does not show any clear gender difference in the 'distance decay' occurring when people live far away from the locations where relevant activities can be carried out. As shown elsewhere (Næss, 2006a, b), there is a tendency towards more frequent visits to cinemas, cafes, rock concerts and discotheques among inner-city residents than among suburbanites, whereas outer-area residents more often than their centrally residing counterparts visit natural areas for recreation. However, the reduction in activity participation with increasing distances to relevant facilities is similar among men and women. For some activities, the frequency of participation varies between men and women (for example, women purchase clothes more frequently than men do, whereas men more often than women participate in team sports and watch sport events). These gender differences are, however, stable across varying residential locations.

In contrast to the findings from several older studies (for example Hjorthol, 1998) we find no gender difference in the proportion of workforce participants among the Copenhagen Metropolitan Area respondents. Yet, male respondents tend to work longer hours than female respondents do (43 hours versus 37 hours on average, respectively). Moreover, while the length of weekly working hours is independent of residential location among men, female respondents tend to work somewhat fewer hours per week if they live far away from downtown Copenhagen ($p = 0.002$). This is in line with findings in Oslo and the surrounding county, where residents of the peripheral parts of the region had somewhat shorter working hours (Hjorthol, 1998). This difference between women and men probably reflects women's larger investment of time in household work and childcare. From a time-budget perspective, this time consumption is likely to be compensated for by shorter working hours and/or shorter commuting time (but also through less leisure time).

Lack of access to a car for daily travel can also, in combination with long distances to relevant facilities, make people give up (or never get involved in) certain activities. In our qualitative interviews we found a few such examples, although these mechanisms do not manifest themselves as identifiable patterns in the statistical material. The clearest example was an interviewee who broke off her education because the distance between her outer-area residence and the university required long and time-consuming journeys. This woman belonged to a household with only one car, and lived in a suburb with quite good public transport connections to downtown Copenhagen. In this family, it was apparently taken for granted that the car 'belonged to the husband', and her occasional car usage took place when the male spouse for some reasons did not need it. Such a distribution of the transport resources within the family is quite common among our survey respondents.

Journeys to work

Commuting distances

In line with findings in a number of previous studies (for example Hanson and Pratt, 1995; Hjorthol, 1998; Lee and MacDonald, 2003), male workforce participants on average work further from their residence (14 km) than female workforce participants do (11.2 km). If median values are considered instead of arithmetic means, the difference is even larger (11 km and 7.2 km, respectively). Also in line with several previous studies, we find longer commuting distances among outer-area residents than among workforce participants living close to downtown Copenhagen. This difference is, however, considerably larger among men than among women, both in absolute and relative figures. As can be seen in Figure 12.2, men living closer than 6 km to downtown Copenhagen have an average commuting distance of 7.5 km, compared to 22 km among their counterparts living more than 22 km away from the Copenhagen city centre. Female workforce participants living less than 6 km from downtown Copenhagen travel on average only slightly shorter distances to reach their workplace than male workforce participants living within the same distance belt do. In contrast, the average commuting distance among female workforce participants living more than 28 km from the Copenhagen city centre is less than two-thirds of the corresponding commuting distance among men.

Thus, among residents of the inner districts of Copenhagen, women choose jobs within the same radius of action as men do. Among residents of the outer suburbs, men's job markets include a considerably larger geographical area than among women. Given the much lower density and range of workplaces in the suburbs than in the inner city, this implies a limitation of suburban women's choices on the job market, compared to men. Among residents living closer than 6 km to the Copenhagen city centre there is no such gender difference.

The limitation of suburban women's job market choices to a smaller geographical area than that of suburban men is probably due to several circumstances. One important reason is the traditional distribution of tasks in the household. Because the female spouse usually spends more time on childcare, cooking, cleaning and daily purchases than her male partner does, less time will be available for wage labour work. This partly results in shorter working hours among women than among men if distances to relevant workplaces are long. The time occupied by wage labour can, however, also be reduced by reducing travel time for journeys to work. The shorter commuting distances among suburban women than among their male counterparts probably reflects such an adaptation. Another case in point is the fact that among couples with one car in the household, the husband is usually considered to be the person with the first right to use the car (see, for example, Hjorthol, 1998, 213–19 for evidence about this from the Oslo region). Since suburban travel is usually faster by car than by public transport, the spouse who does not have access to a car must

– if thresholds for acceptable travel time are equal – work closer to home than the car-commuting spouse.

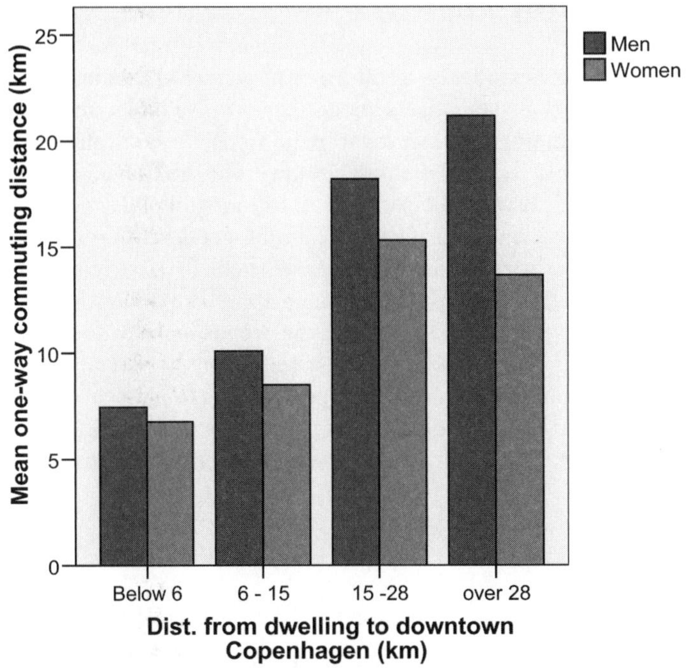

Figure 12.2 Average total travel distance Monday-Tuesday among female and male respondents living within different distance intervals from downtown Copenhagen
N = 933 women and 858 men

A closer look at the data (cf. Table 12.1) indicates that the location of the dwelling relative to downtown Copenhagen exerts a very strong influence on the commuting distances among men when controlling for a number of demographic, socio-economic and attitudinal potential factors of influence. In addition, the density of jobs and inhabitants in the local area surrounding the dwelling appears to influence commutes slightly between both sexes. Among women, commuting distances also tend to be influenced by the location of the dwelling relative to the closest second-order centre. The latter is a further indication of women's more local job market orientation compared to men.

Surprisingly, the difference between men and women in average commuting distances is nearly twice as large among two-car couples than among one-car couples. This suggests that the purchase of a second car in the household enables men to commute longer (as they no longer need to negotiate with their wife about the use of the car for commuting), whereas women tend to use car ownership in terms of saving travel time.[5]

Travel modes for journeys to work

Among our suburban respondents, we find a clear gender difference in travel modes for journeys to work, suggesting that a gender-based inequality in the right to use the family's car for commuting is also present in the Copenhagen Metropolitan Area.

Among respondents living more than 28 km from the Copenhagen city centre, 90% of the male workforce participants drive to the workplace, compared to only 50% of the female workers (Figure 12.3). A similar gender difference exists among residents of the inner suburbs, although the proportions of car drivers are lower both among women and men.[6] Interestingly, among those respondents who live closer than 6 km to the city centre of Copenhagen, the proportion of women who drive by car to the workplace is higher than the corresponding proportion among men. In this distance belt, the availability of local job opportunities is so high and the advantages of travelling by car so small that only one-quarter of the male workforce participants drive to their workplace. Among female workforce participants living in the same distance belt, almost one-third drives by car.

5 Apart from the three urban structural variables and car ownership, the commuting distances of females tend to be higher if she has a high income and if the respondent has moved to her present dwelling less than five years ago. The effect of income probably reflects both the fact that high-income jobs are usually more specialised than low-income jobs and therefore less likely to be found locally, and the higher mobility facilitated by a high income. The effect of having moved recently is interesting, as no similar effect is found among the male respondents. To the extent that concerns about the distance to the workplace is at all emphasised when choosing a new place to live, the effect of having moved may suggest that limiting men's commuting distance is given higher priority than limiting the distance to the woman's workplace when couples move. Although this explanation must be considered quite speculative, it fits with the data. Among the male respondents, the following demographic and socio-economic characteristics appear to contribute to increased commuting distances the most: Being a part-time pensioner and few schoolchildren in the household. The effect of schoolchildren suggests that fathers tend to avoid long commutes when there are children between 7 and 18 years in the household. This may appear somewhat surprising as no similar effect is found among female respondents. But it could also be turned the other way round: When the children have grown up or are so small that they do not involve so much parent participation in connection with leisure activities in the afternoon, men increase their geographical job market while women stick to their more local orientation, possibly due to other household chores. The effect of being a part-time male pensioner is also surprising – perhaps a reduced number of workdays makes it more attractive to move to a peripheral dwelling or a summer cottage far away from the city centre?

6 There is a much smaller gender difference in the proportion of car drivers among commuters living in the third distance belt (15–28 km from downtown Copenhagen) than in the second and fourth.

Table 12.1 Results from a multivariate analysis of the influence of various independent variables on the daily one-way commuting distance (km) of workforce-participating respondents

	Women[1]			Men[2]		
	Regression coefficients		Level of significance (p values, two-tail)	Regression coefficients		Level of significance (p values, two-tail)
	Unstandardised (B)	Standardised (Beta)		Unstandardised (B)	Standardised (Beta)	
Location of the residence relative to downtown Copenhagen (non-linear distance function, values ranging from 0.66 to 3.80)	1.613	0.197	0.002	3.880	0.405	0.000
Personal annual income (1000 DKK)	0.0096	0.139	0.002	0.0043	0.094	0.048
Has moved to the present dwelling less than five years ago (yes = 1, no = 0)	2.923	0.137	0.001	0.937	0.037	0.396
Logarithm of the distance from the residence to the closest second-order urban centre (values ranging from 2.49 to 4.46)	2.784	0.097	0.104	-1.608	- 0.048	0.407
Density of inhabitants and workplaces within the local area of the residence (inhabitants + workplaces per hectare)	- 0.0147	- 0.093	0.153	- 0.020	- 0.106	0.108
Number of household members below 7 years of age	- 1.577	- 0.092	0.085	0.048	0.003	0.958
Regular transport of children to school or kindergarten (yes = 1, no = 0)	2.404	0.092	0.086	2.226	0.064	0.183
Being a student (yes = 1, no = 0)	- 2.469	- 0.061	0.174	0.538	0.010	0.812
Extensive technical or economic education (yes = 1, no = 0)	2.328	0.058	0.156	1.792	0.059	0.177
Possession of a driving license (yes = 1, no = 0)	- 1.767	- 0.056	0.200	- 0.263	-0.005	0.913
Index for transport attitudes (high value = car-oriented attitude, values ranging from -17 to 11)	0.101	0.052	0.305	0.027	0.013	0.815
Short or medium-length education as a tradesman or industrial worker (yes = 1, no = 0)	- 2.343	-0.050	0.222	1.740	0.059	0.183
Age (deviation from being 'middle-aged', logarithmically measured)	- 1.487	- 0.045	0.328	- 2.508	- 0.019	0.688
Number of cars per adult household member	1.054	0.038	0.447	1.338	0.041	0.415
Number of household members aged 7-17	- 0.443	- 0.034	0.453	- 1.490	- 0.105	0.025
Logarithm of the distance from the residence to the closest urban rail station (values ranging from 1.90 to 4.47)	0.547	0.028	0.581	- 0.188	- 0.008	0.864
Index for environmental attitudes (high value = much concerned about the environment, values ranging from -17 to 11)	0.052	0.025	0.618	0.152	0.067	0.211
Being a part-time pensioner (yes = 1, no = 0)	0.685	0.006	0.672	-8.981	- 0.098	0.021
Constant	- 4.086		0.506	11.181		0.137

Notes To Table 12.1

N = 535 female and 469 male respondents from 29 residential areas in the Copenhagen Metropolitan Area. Adjusted $R^2 = 0.174$ (women) and 0.221 (men). The variables have been ordered according to the absolute values of their standardised regression coefficients among female respondents.

[1] If only variables meeting a required significance level of 0.15 are allowed to be included in the model, only the following variables show significant effects on the commuting distances of female respondents: Location of the residence relative to downtown Copenhagen ($p = 0.000$), personal income ($p = 0.000$), having moved to the present residence less than five years ago ($p = 0.001$), Logarithm of distance from residence to closest second-order centre ($p = 0.022$), and local-area density ($p = 0.137$).
[2] If only variables meeting a required significance level of 0.15 are allowed to be included in the model, only the following variables show significant effects on the commuting distances of male respondents: Location of the residence relative to downtown Copenhagen ($p = 0.000$), being a part-time pensioner ($p = 0.007$), number of household members aged 7–17 ($p = 0.010$), personal income ($p = 0.024$), and regular transport of children to school or kindergarten ($p = 0.067$).

Apparently, among residents living in locations where the use of car for commuting is considered to be the fastest and most attractive alternative, men have the privilege of using the car if the household has only one such vehicle at its disposal. In situations where car commuting is not considered to be so attractive, the husband typically leaves the car at home, thus making it available for the wife. If she – as is the case among women more often than among men – has errands to carry out during the journey to or from the workplace (for example bringing children to/from kindergarten), she may choose to use the car.

Shopping and leisure trips

Use of local shopping and leisure facilities

Female respondents in the Copenhagen Metropolitan Area investigation tend to use local facilities slightly more often than men do when purchasing daily necessities as well as books and CDs. These differences do not vary in any systematic way with the location of the residence. However, when visiting cafes and cinemas, women in the suburbs tend to use local facilities to a somewhat greater extent than men do, while the opposite is the case among respondents living less than 6 km away from downtown Copenhagen. This appears to reflect a higher tendency among men than among women to visit cafes and cinemas in the city centre. Possibly, men more often than women go to see a film or visit a café after work, if such facilities are available in the proximity of the workplace. Given the strong concentration of workplaces in the inner and central part of Copenhagen, this might lead to a higher propensity for using local cinemas and cafes among men than among women living in the central parts of the region, and a correspondingly lower tendency among men than among women living in the outer districts.

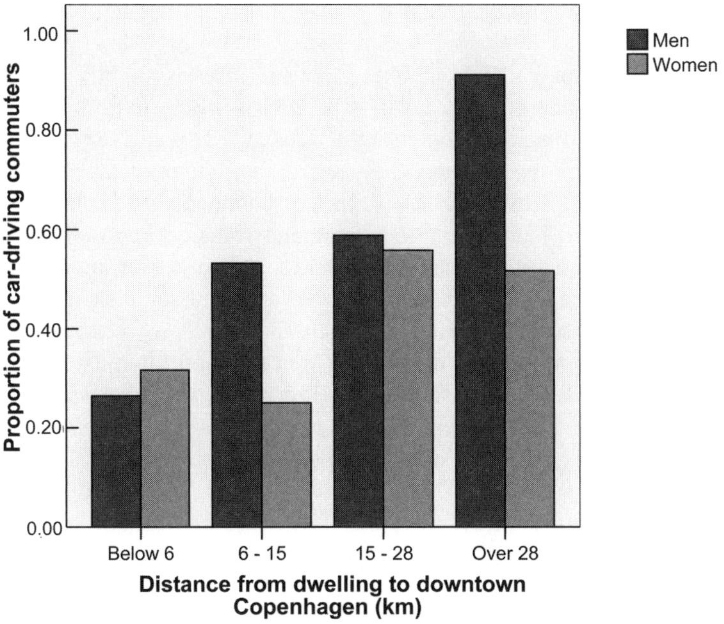

Figure 12.3 Proportions of car-driving commuters among female and male respondents living within different distance intervals from downtown Copenhagen
N = 632 women and 546 men

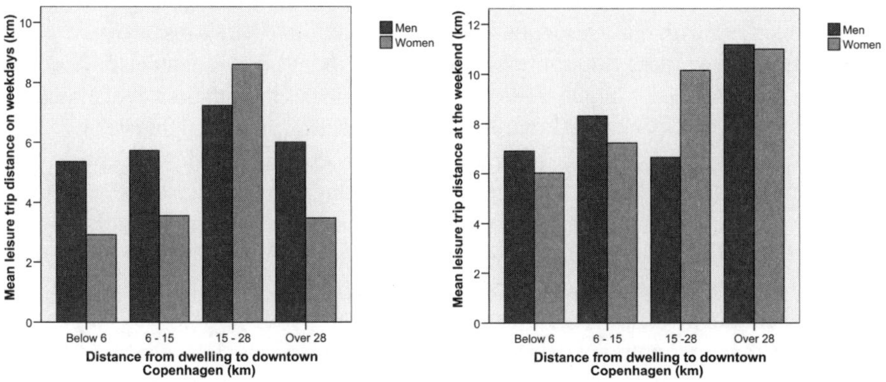

Figure 12.4 Mean trip lengths of leisure trips on weekdays (to the left) and on the weekend (to the right) among female and male respondents living within different distance intervals from downtown Copenhagen
N = 115 women and 132 men

Leisure trip frequencies and distances

The proportion of trips carried out for leisure purposes is slightly higher among women than among men on weekdays (10% and 8%, respectively), but somewhat higher among men than among women at the weekend (21% and 18%, respectively). The small difference during weekdays may mirror the shorter average working hours of female respondents, leaving a little bit more time available for leisure trips. At the weekend, activities like cooking, cleaning and visits occupy more time among women than among men, thus enabling men to make somewhat more leisure trips than women on Saturday and Sunday. There is, however, some spatial variation in the frequencies of leisure trips among women and men. On weekdays, the higher number of leisure trips among women than men first and foremost occurs among respondents living closer than 6 km to downtown Copenhagen. Conversely, the lower number of weekend leisure trips among women than among men is a phenomenon primarily occurring among respondents living more than 28 km away from downtown Copenhagen. These differences may indicate that, within the tight time budgets of weekdays, the possibility of using the time made available by women's shorter working hours by making leisure trips is first and foremost found among those women who live close to a concentration of facilities such as restaurants, cafes, cinemas, theatres and gyms, typically found in the inner parts of the metropolitan area. With the more relaxed time budget at the weekend, distance decay in visiting these facilities seems to occur only among residents living on the metropolitan periphery, in particular those whose time available for 'pure' leisure activities is limited by traditional female tasks like cooking, cleaning and visiting elderly relatives.

The length of leisure trips shows only small gender variations at the weekend. On weekdays, however, there is a considerable variation in leisure trip lengths between men and women, with male respondents typically making leisure trips nearly twice as long as those of women. Both on weekdays and at the weekend, gender differences in leisure trip lengths are largely independent of the location of the residence relative to the Copenhagen city centre (Figure 12.4). Yet, respondents living between 15 and 28 km from downtown Copenhagen represent an exception. Among residents in this distance belt, women's leisure trips are on average longer than men's on weekdays as well as at the weekend. Women's leisure trips appear to be directed towards 'urban' facilities like cafes and cinemas to a somewhat stronger extent than among men. The provision of a number of service facilities is particularly low within this distance belt – even lower than in the most remote of the four distance belts, where the competition from facilities in the central parts of the region is smaller and there is hence a stronger base for local facilities than in the not-so-remote suburbs. The leisure trips of females living in the third distance belt therefore tend to be quite long. Apparently, men make use of suburban and rural leisure facilities to a greater extent, such as football grounds, recreational forests, golf courses etc. For residents in the third distance belt, such facilities will usually be available not too far from the dwelling.

Total travel over the weekdays

Travelling distances

Having seen that average commuting distances differ between male and female respondents of the Copenhagen Metropolitan Area study, particularly among those who live in the suburbs, whereas the main gender difference in the influence of residential location on leisure trips is a higher frequency of such trips among women than men living close to downtown Copenhagen, let us now turn to the total travelling distance over the five weekdays (Monday–Friday). We should also bear in mind that both commuting distances and leisure trip distances are on average shorter among women than among men. We could therefore expect the gender difference in total distances travelled over the weekdays to be greater among outer-area than among inner-city respondents.

Below, we will first take a look at the relationships between residential location and travel distances among men and women who are all workforce participants. Thereupon, the situation among non-participants of the workforce will be addressed.

Figure 12.5 shows how the total travelling distance over weekdays among working men and women varies with the distance belt from the Copenhagen city centre within which the respondents live. In the inner distance belt, women and men's travelling distances are almost the same. In the suburbs, and in particular the two outer distance belts, there is a considerable difference between men and women in the distances travelled. Women limit their radius of action, compared to men, when they live in outer parts of the metropolitan area. Given the lower concentration of facilities – workplaces, shops as well as leisure opportunities – in the suburbs than in the city centre, the smaller radius of action among women implies that their choices both on the job market and regarding leisure opportunities are limited, compared to men. The most common response among men to the low local provision of facilities in the outer areas thus seems to be increased travel, whereas women's response to these conditions seems to be reduced activity participation and/or limitation of choices among different opportunities for performing an activity. Distinct from that, the radius of action on weekdays appears to be equal among women and men living less than 6 km from downtown Copenhagen. From a gender equality perspective, women living in the inner districts of the metropolitan area thus seem to be in a better position than their outer-area sisters.

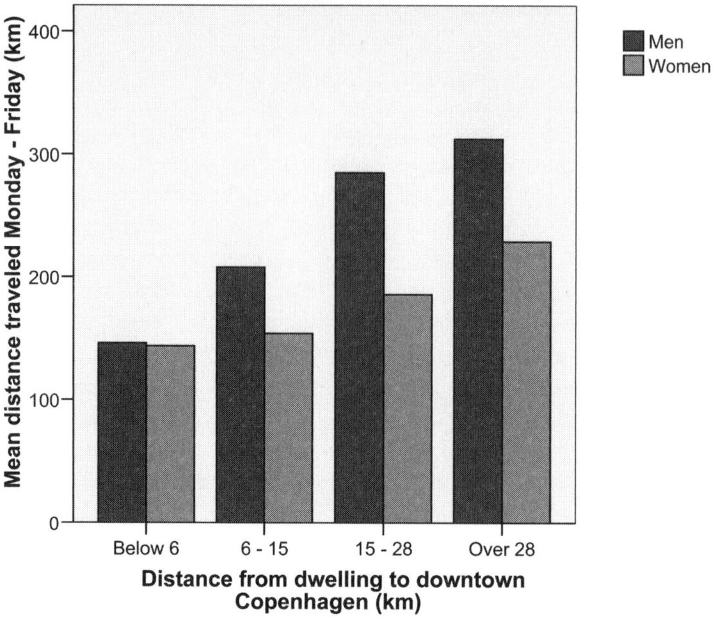

Figure 12.5 Average total travel distance Monday-Tuesday among female and male workforce participants living within different distance intervals from downtown Copenhagen N = 632 women and 546 men

Table 12.2 compares the statistical relationships of different potential factors of influence on the travelling distances of women and men. Among women, the distance travelled on weekdays appears to be influenced by the location of the dwelling relative both to the closest urban rail station (where a local centre usually exists), the closest second-order centre and the main centre of Copenhagen, with the former effect the strongest of the three. In contrast, the location of the residence relative to downtown Copenhagen is the only urban structural variable showing a significant effect on the distance travelled by men on weekdays. This illustrates that the more local orientation of women than of men is not confined to choices of workplaces, but applies to weekday travel in general. Among workforce participant women who live far from the Copenhagen city centre, the travelling distance on weekdays may still be moderate if the dwelling is located close to a local centre. Among working men living on the periphery, proximity to a local centre does not seem to affect weekday travelling distances, a fact indicating a much lower propensity to use local facilities among men than among women.

The commute still makes up a high proportion of the daily travel of workforce participants, in particular those living in the outer areas of the metropolitan area. Taking a look at the distances workforce participants travel on weekdays for other purposes than commuting, we find that men's non-work travelling distances are longer than women's in all the three outer distance belts, whereas working women living closer than 6 km to downtown Copenhagen travel a little longer for non-work purposes than men living within the same distance belt. Among working women, the distance travelled for non-work purposes tends to increase, the further away they live from the closest local centre, but increases the higher the density of the local area around the dwelling. Increased leisure travel in order to compensate for lack of greenery in the local area might be an explanation for the latter effect (see Næss, 2006c for a more in-depth account on the phenomenon of 'compensatory travel'). Possibly, women living in low-density areas also reduce the number of non-work trips due to the few 'urban' leisure facilities within reach.

Among those who are not participants in the workforce, we find – similar to the situation among workforce participants – that gender differences in travelling distances over the weekdays only exist in the three outer of the four distance belts. Among those respondents living closer than 6 km to downtown Copenhagen, women travel on average slightly longer distances than men do, but this difference is very small. Overall, non-workers travel considerably shorter distances over the weekdays than workforce participants do.

A more detailed analysis reveals that non-working men's travelling distances tend to increase the further away from downtown Copenhagen they live and the lower the density of the local area around the dwelling. Since high-density areas are mainly located in the central parts of the region, this implies that non-working men's travelling distances tend to increase the more peripheral their dwelling. This resembles what was found among male workforce participants. Among female non-workers, travelling distances over the weekdays do not seem to be influenced by any of the urban structural variables. In the absence of a daily commute, women are even more locally oriented. Since neither proximity to local nor higher-order centres seems to affect the travel behaviour among this group, non-working women living far away from the various types of centres logically must make fewer trips to the respective centre categories than those living close to the centres do. Again, we see the same pattern as in the analyses mentioned earlier: Women living peripherally tend to limit their use of and choice among facilities, while men compensate for lack of proximity to facilities by travelling more.

Table 12.2 Results from a multivariate analysis of the influence of various independent variables on the total distance travelled (km) over the weekdays (Monday–Friday) among workforce-participating respondents

	Women[1]			Men[2]		
	Regression coefficients		Level of significance (p values, two-tail)	Regression coefficients		Level of significance (p values, two-tail)
	Unstandardised (B)	Standardised (Beta)		Unstandardised (B)	Standardised (Beta)	
Occupational trips during the investigated week (yes = 1, no = 0)	95.468	0.225	0.000	92.299	0.225	0.000
Number of cars per adult household member	71.457	0.173	0.000	89.266	0.185	0.001
Number of household members below 7 years of age	- 40.007	- 0.156	0.002	- 11.757	- 0.036	0.433
Logarithm of the distance from the residence to the closest urban rail station (values ranging from 1.90 to 4.47)	37.431	0.133	0.007	20.018	0.054	0.255
Index for transport attitudes (high value = car-oriented attitude, values ranging from -17 to 11)	3.212	0.112	0.015	3.347	0.100	0.066
Logarithm of the distance from the residence to the closest second-order urban centre (values ranging from 2.49 to 4.46)	43.445	0.101	0.104	-13.227	- 0.024	0.674
Location of the residence relative to downtown Copenhagen (non-linear distance function, values ranging from 0.66 to 3.80)	11.489	0.093	0.113	34.016	0.213	0.000
Overnight stays away from home more than three nights during the investigated week (yes = 1, no = 0)	59.523	0.089	0.019	54.564	0.073	0.082
Has moved to the present dwelling less than five years ago (yes = 1, no = 0)	22.352	0.070	0.072	5.201	0.012	0.769
Personal annual income (1000 DKK)	0.054	0.058	0.164	0.027	0.037	0.415
Number of days of appearance at the workplace during the investigated week	5.592	0.053	0.166	- 4.750	- 0.034	0.420
Density of inhabitants and workplaces within the local area of the residence (inhabitants + workplaces per hectare)	0.125	0.052	0.400	- 0.142	- 0.044	0.481
Index for environmental attitudes (high value = much concerned about the environment, values ranging from -17 to 11)	- 1.501	- 0.048	0.289	2.918	0.077	0.134
Extensive technical or economic education (yes = 1, no = 0)	29.324	0.047	0.217	56.643	0.115	0.006
Being a student (yes = 1, no = 0)	- 22.089	- 0.036	0.370	15.828	0.019	0.854
Regular transport of children to school or kindergarten (yes = 1, no = 0)	13.104	0.033	0.505	- 18.109	- 0.030	0.510
Number of household members aged 7-17	- 5.954	- 0.030	0.458	- 13.980	- 0.059	0.189
Being a pensioner (yes = 1, no = 0)	- 29.000	- 0.021	0.583	49.339	0.029	0.469
Possession of a driving license (yes = 1, no = 0)	- 8.531	- 0.018	0.658	- 13.821	-0.016	0.708
Age (deviation from being 'middle-aged', logarithmically measured)	2.584	0.005	0.901	- 53.201	- 0.080	0.081
Short or medium-length education as a tradesman or industrial worker (yes = 1, no = 0)	- 2.278	-0.003	0.931	31.631	0.085	0.122
Constant	- 204.391		0.019	152.695		0.215

Notes to Table 12.2

N = 571 female and 513 male respondents from 29 residential areas in Copenhagen Metropolitan Area. Adjusted $R^2 = 0.247$ (women) and 0.229 (men). The variables have been ordered according to the absolute values of their standardized regression coefficients among female respondents.

[1] If only variables meeting a required significance level of 0.15 are allowed to be included in the model, only the following variables show significant effects on the commuting distances of female respondents: Occupational trips during the investigated week (p = 0.000), number of cars per adult household member (p = 0.000), location of the residence relative to the closest urban rail station (p = 0.005), index for transport attitudes (p = 0.001), number of household members below 7 years of age (p = 0.001), overnight stays away from home (p = 0.017), location of the residence relative to the closest second-order centre (p = 0.111), personal income (p = 0.047), having moved to the present residence less than five years ago (p = 0.057), location of the residence relative to downtown Copenhagen (p = 0.131), and number of days of appearance at the workplace (p = 0.115).

[2] If only variables meeting a required significance level of 0.15 are allowed to be included in the model, only the following variables show significant effects on the commuting distances of male respondents: Location of the residence relative to downtown Copenhagen (p = 0.000), occupational trips during the investigated week (p = 0.000), number of cars per adult household member (p = 0.000), extensive technical or economic education (p = 0.004), index for transport attitudes (p = 0.065), overnight stays away from home (p = 0.025), index for environmental attitudes (p = 0.136), and short or medium-long education as a tradesman or industrial worker (p = 0.116).

Travel modes

Considering all types of travel purposes together, we find clearer gender differences in travel modes on weekdays than when considering journeys to work only. For both sexes, residents of outer areas tend to have higher shares of car travel and lower shares of non-motorised modes than their counterparts living closer to the Copenhagen city centre. Regardless of the location of the dwelling relative to downtown Copenhagen, working women have a lower proportion of car travel and a higher proportion of travel by bike or on foot than among male workforce participants (Figure 12.6). The difference between men and women in car usage is lower in the innermost and the outermost distance belt, compared to the two distance belts in-between. This corresponds to a clear difference in the use of public transport between men and women living in the latter distance belts. Among residents of the innermost and outermost distance belt, the shares of public transport are the same among women and men.

As shown earlier, the advantages of using cars (notably in terms of travel time) are much smaller when travelling in the inner city than on the outskirts of the metropolitan area, and the need for using motorised travel is lower than in other parts of the region. The competition between the spouses of one-car households for car use will therefore probably not be so strong. At the same time, public transport services are usually very poor in the most peripheral parts of the region, and many families have two cars. This implies that the combination of a widespread need for motorised travel and a moderate frequency of two-car households is most likely to occur among residents in the second and third distance belt. In such situations, one spouse is likely to take the car while the other spouse goes by public transport (unless they are travelling together). The much higher share of public transport among female than among male workforce participants living within these distance belts indicates that the family car is typically used by the husband, while the wife travels by public transport.

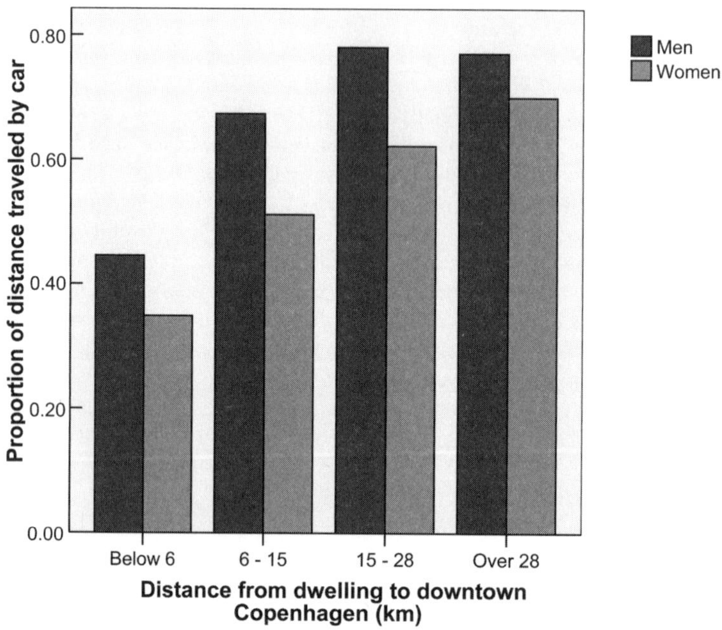

Figure 12.6 Proportions of total travel distance Monday-Tuesday travelled by car among female and male workforce participants living within different distance intervals from downtown Copenhagen
N = 632 women and 546 men

Among non-participants in the workforce, the influence of residential location on transport modes as well as gender differences is largely similar to those among workforce participants (but with generally lower shares of car travel). The proportions of non-motorised travel are, however, about equally high among non-working men and women (but still with considerably higher shares in central than in peripheral parts of the region). Probably, this is because non-working men do not need to make the long commutes that are typical among many male workforce participants, and thus have a higher number of destinations within acceptable biking distance.

Concluding remarks

In the introduction to this chapter, we mentioned the 'feminine' and 'masculine' urban models discussed among Norwegian transport researchers in the 1980s. The Copenhagen Metropolitan Area study (and indirectly also a number of other studies showing how urban structural conditions affect accessibility, car dependency and travel) supports the assumptions upon which the ideas about the 'feminine' and 'masculine' city models were constructed. Moreover, the study shows that in the present urban structure of the Copenhagen region, women's accessibility is more equal to that of men if they live in the inner districts. Among suburbanites, there is a clear tendency for women to find it necessary to choose among a narrower range of job opportunities and leisure facilities than their male counterparts.[7]

In the current discourse on sustainable urban development, many debaters appear to consider the transport consequences of land use and transportation infrastructure investments as a typical 'boys' topic', perhaps because so many of the analyses carried out within this field have been dominated by the 'number crunching' of traffic engineers. At least in Denmark, contemporary feminist utopias depict ecologically sustainable cities as small local communities situated in green surroundings, with local food production and a local balance between dwellings and workplaces. *Gemeinschaft* and *life world* are the focus rather than *gesellschaft* and *system world.* However, the transportation implications of such a settlement structure – given that the residents are not to be prohibited from using facilities outside their little eco-village – is seldom discussed. The attempts made in the 1980s to open a gender equality debate within the field of transportation was thus indeed pioneering, as the 'feminine' urban model emphasised proximity between different city facilities as a strategy where the inhabitants would not have to choose between a substantial car dependency and constrained opportunities for choice. Both from a feminist and sustainability perspective, such an urban developmental path should be encouraged.

References

Hanson, S. and Pratt, G. (1995), *Gender, Work and Space* (London: Routledge).

Hjorthol, R. (1998), Hverdagslivets reiser. En analyse av kvinners og menns daglige reiser i Oslo (The travels of everyday life. An analysis of daily trips among women and men in Oslo), TØI rapport 391/1998 (Oslo: Transportøkonomisk institutt).

Jenseth, S. et al. (1986), Likestiling i samferdsel (Gender equality within transportation) (Oslo: Transportøkonomisk institutt).

Jørgensen, G. (1992), Erhverv i boligkvarteret - en vej til bedre bymiljø? (Workplaces in the residential area – a strategy for a better urban environment?) (Hørsholm: Statens Byggeforskningsinstitut).

7 Similar gender differences in the influence of residential location on commuting distances and daily travel as in the Copenhagen study were found in a recent study in Hangzhou Metropolitan Area in China (Næss, 2007).

Lee, B.S. and McDonald, J. (2003), 'Determinants of Commuting Time and Distance for Seoul Residents: The Impact of Family Status on the Commuting of Women', *Urban Studies* 40 (7), 1283–302.

Næss, P. (2006a), *Urban Structure Matters. Residential Location, Car Dependency and Travel Behaviour* (London/New York: Routledge).

Næss, P. (2006b), 'Accessibility, Activity Participation and Location of Activities. Exploring the Links Between Residential Location and Travel Behavior', *Urban Studies* 43 (3) 627–52.

Næss, P. (2006c), 'Are Short Daily Trips Compensated by Higher Leisure Mobility?' *Environment & Planning B* 33, 197–220.

Næss, P. (2007), Residential Location and Travel in Hangzhou Metropolitan Area, NIBR report 2007:1 (Oslo: Norwegian Institute for Urban and Regional Research).

Chapter 13

Daily Mobility of Men and Women – A Barometer of Gender Equality?

Randi Hjorthol

Introduction

Our present-day society is characterised by speed, a hectic life and mobility, both physical and virtual (Lash and Urry, 1994, 252; Urry, 2000, chapter 3). The development of urban areas from the middle of the 19th century to the present day has to a great degree been characterised by large transport infrastructure projects designed to increase mobility and accessibility. To travel is thus an important part of modern life. Travel is associated with freedom, and to travel where and whenever there is a need or wish is seen to be a basic right in a democratic society. The ability to travel is a precondition for participating in different arenas in society. Most people work in a place other than where they live, and activities that earlier were located in the home or the neighbourhood are now spread over large geographical areas. In this situation, organising everyday life very often demands travel to conduct the various necessary activities. Urban sprawl has scattered everyday activities over a large area, which forces people to spend a considerable amount of time on the move. Rapid transport means moving people over long distances in a short time. And information and communications technology (ICT) has increased our ability to communicate; the world has 'shrunk'. Harvey's (1989, 147) concept of *time-space compression* highlights the same phenomenon, and underscores the ability (and need) to move quickly in time and space. Doreen Massey (1994, 149) has pointed out the fact that this will not be true for all social groups; both accessibility and mobility vary. The gendered aspects of this variation, its development over time, and the consequences of the variations are the focus of this chapter.

The relationship between gender, transport and daily physical mobility has attracted considerable attention in recent years. Previous studies have documented differences between men's and women's travel patterns and access to transport resources. Women have had poorer access to private transport resources than men, they travel shorter distances and their travel purposes are different. The environmental consequences of the differences between the genders have also gained attention. In general, men's travel patterns have caused more environmental problems than women's, because they travel longer distances and use cars more often than women. This raises the questions of whether women and men's daily travel patterns are converging, what will happen to women's travel pattern when men and women's everyday life become more alike (if they do) and what will the consequences be?

To examine these questions I will present an analysis of everyday travel and mobility in Norway from 1992 to 2005. Everyday travel and mobility reflect social activities and positions and are related to temporal and spatial aspects of social institutions, the division of labour and unpaid work, and access to resources. My point of departure is therefore to examine the development of gendered aspects of important societal areas related to paid and unpaid work as a condition for the daily travel of both men and women. I will incorporate several factors suggested as a framework for understanding of gender and mobility by Law (1999, 576).

Before analysing these gendered aspects and the development of everyday travel, I will provide a brief definition of everyday mobility and an overview of previous research.

What is everyday travel?

Very simply, there are three reasons why people travel in their daily lives:

- Because they *want to* – they want to change place for some reason, to meet other people, show off their new car, or ride a bike in the fine weather.
- Because they *can* – they have access to time, money, to a transport system and/or private transport means.
- Because they *have to* – the spatial organisation of social activities demands travelling, and different obligations 'demand' travelling.

These three reasons are often interrelated. A significant percentage of daily activities that contribute to one's health and well-being, from work to leisure activities to education, presuppose that one will travel. How far people travel, how often and the type of transport mode used depends upon both individual and social conditions.[1]

John Urry (2003, 163) claims that travel 'results in intermittent moments of *physical proximity*' which are *obligatory, appropriate or desirable*. He claims that proximity occurs for five different reasons:

1. *Legal, economic and familial obligations* – for example travel to work, family events, school, hospital.
2. *Social obligations* – less formal demands, but very often involving very strong normative expectations such as physical presence and attention. Such social obligations involve seeing the other individual face-to-face, are associated with 'quality time', and are often in specific locations involving long travel.
3. *Object obligations* – including the need to be present to sign contracts or to work on or to see various objects (mostly related to work).

1 The interaction between virtual and physical mobility will also have an impact on organising everyday life. The use of the mobile phone, in particular, seems to be important for the travel patterns in families with children (Hjorthol et al., 2006). This subject will not be discussed in this chapter.

4. *Obligation to place* – to sense a place or a certain kind of place directly – walking within a city, being at the seaside, walking in the mountains (adventure).
5. *Event obligations* – to experience a particular life event.

Daily mobility comprises all five obligations, but it is perhaps the first two of them that are the most common in everyday life. I think it is important to underscore the significance of the temporal and spatial aspects of these obligations. The temporal and spatial structures of society are both the medium and the outcome of social actions, which also means that spatial structures are not neutral – they are carriers of meaning, control and power. With respect to gender, time and space are of great importance, because the organisation of daily activities, mobility and travel are linked to the gendered distribution of work and household activities, access to resources and the construction of identity. Power is exercised in each of these.

Previous studies

The gendered aspects of travel behaviour have attracted considerable attention for roughly three decades. Initially, most studies focused on commuting. Findings showed that employed women tended to have shorter work trips than men and used cars less (Dasgupta, 1983; Hanson and Pratt, 1988; Hanson and Johnston, 1985; Madden 1981). These findings were explained by invoking various factors. Traditional gender roles and women's dual role was often used both as an explanation of women's shorter working hours and shorter distances to work; these factors together assign more responsibility for household maintenance to women than to men (Ericksen, 1977; Faganini, 1987; Hanson and Pratt, 1988; Hanson and Johnston, 1985). Some studies indicated that both the number of children (Faganini, 1987; Hanson and Pratt, 1995) and their age (Ericksen, 1977) have an impact on the distance that female employees commute to work. Women with small children or many children had shorter journeys to work than other women. Madden (1981) found that married women with children had the shortest journeys to work, also when controlled for the effects of income, labour force and housing market characteristics. A study by Manning (1982) found the same result. Another study showed that marital status was the reason for the shorter journeys to work (Johnston-Anumonwo, 1992). In other words, there was evidence that household responsibilities or division of labour within the family played a significant role in women's choice of both working hours and work location.

Several studies have suggested that length of the journey to work varied with *occupational status*, that is, high-status workers travelled longer distances to work than low-status workers (Sirmans, 1977; Gera and Kuhn, 1978). These differences also held for women (Andrews, 1978; Faganini, 1987; Hanson and Pratt, 1988; Villeneuve and Rose, 1988). But even if there were differences between occupational groups of women, there seemed to be significant differences between men and women in the same occupational groups (Hanson and Johnston, 1985).

The differences between men and women in terms of length of journey to work have also been said to reflect the *spatial distribution* of their respective employment opportunities. Hanson and Johnston (1985) summarised the conclusions of some of these studies, which are concentrated mostly in urban areas.

Traditionally, men have had *better access to a car* than women and have used it more for work-related travel (Guiliano, 1979; Hanson and Johnston, 1985; Madden and White, 1979; Rutherford and Wekerle, 1988). Lack of suitable means of transport might restrict women's choice in the labour market, and so their greater dependency on public transportation than that of men may explain their shorter journey to work or their situation as 'captive riders' (Rutherford and Wekerle, 1988). In general (at least European) women have had less access to a car, especially elderly women (OECD /ECMT, 2000; OECD, 2001; Hjorthol and Sagberg, 1998; Rosenbloom, 2002).

In addition to research related to the differences in commuting, studies also show that women make more household and family support trips (Hanson and Hanson, 1980; Hanson and Johnston, 1985; Hjorthol, 1998, Chapter 7; Rosenbloom, 1987), and that women make fewer recreational trips.

Men and women also perceive the car and public transport differently (Hjorthol, 2002a). Women have a more positive relationship with public transport than men, while the car belongs more to the masculine than the feminine domain (Wachs, 1992). The car as a masculine object has also been supported by the results of Scandinavian research, which has focused on car use, marketing, car club membership and driving as an expression of a form of mastery (Hagman, 1995; Lamvik, 1994; Rosengren, 1993).

Current interest in research on the relationship between gender and mobility in everyday life seems to have declined, compared to the 1970s and the 1980s. Do planners and researchers believe that these differences have disappeared? The newer (but relatively few) studies of gendered aspects of transport indicate that there are still significant differences, which should be of interest for contemporary land use and transport planning. A study of commuting and gender in Italy indicates that women still have shorter work trips than men and that they use public transport more than men (Cristaldi, 2005). The study also indicates, however, that there are significant differences between groups of women. Well educated women travel longer than women with less education. A Canadian study from Quebec indicates social change; women have switched from public transit to driving cars to commute (Vandermissen et al., 2003). But studies indicate that women still experience more spatial and time budget constraints than men (for example Kwan, 2000). Some researchers claim that access to private transport (car) is the key factor in determining women's mobility and economic inclusion (Dobbs, 2005).

The analysis presented in this chapter is an attempt to provide a social context for the social changes or lack of changes that can be observed in the transportation field. A comprehensive approach is employed in an analysis of daily travel. Changes in daily travel and mobility reflect social changes in other fields. But changes in daily travel and mobility will also have an impact on other areas of society, which will be discussed in the final section.

Gender relations

Gender relations are socially and culturally constructed and as such vary in time and space. These constructions are reflected in several social areas. Here I will focus on three, which traditionally have been highly gendered and which are related to daily travel and mobility.

- Labour market.
- Household work.
- Responsibility for children and elderly relatives.

On average Norwegian women have been paid less than men, while more of them are occupied in health and care work. They also have fewer fringe benefits and shorter working hours (Statistics Norway http://www.ssb.no/norge/inntekt/).

Table 13.1 shows time use for Norwegian men and women in the age group 25–44 years from 1971 to 2000. I have chosen to present data from 1971 to 2000 to provide long-term trends. Women spend more of their time doing household work such as cooking, washing and shopping than men. They also use more of their time caring for children than men do.

Table 13.1 Time use per day for men and women 25–44 years (Norway 1971, 1980, 1990, 2000, hours and minutes)

Activities	1971		1980		1990		2000	
	Men	Women	Men	Women	Men	Women	Men	Women
Paid work	5.32	1.44	5.10	2.25	4.51	2.53	5.05	3.18
House work	0.33	4.33	0.43	3.09	0.42	2.25	0.48	1.52
Private care work	0.30	1.18	0.40	1.22	0.48	1.32	0.39	1.12
Shopping	0.15	0.24	0.19	0.28	0.17	0.27	0.19	0.27

Source: Vaage, 2002.

Table 13.1 shows several interesting trends. While women have nearly doubled their hours of paid work, men's labour input has been slightly reduced. But men continue to work longer hours than women. The hours devoted to housework have decreased dramatically during this 30-year period. Men have increased their contribution to the household duties by 15 minutes in 30 years. Women still care for children and elderly people more than men, and time use has changed relatively little for both men and women during this period. For men there has been a minor increase. The same applies for shopping, which also is mostly carried out by women.

These figures indicate that even if there has been a radical change in women's relation to paid work and to household work, men have assumed only slightly more of the household responsibilities. These time use changes also reflect cultural changes. The norms of housework have evidently changed, along with the greater ownership of labour saving machines such as dishwashers and washing machines (but the clothes are washed more frequently). More precooked meals and some dust under the sofa are apparently acceptable.

But the norms related to parenting or mothering have changed little, when time use is the measurement. The focus on children in Norwegian society is significant. Giving children 'quality time', engagement in their leisure activities and providing the best opportunities are examples. In western cities there has been an increasing involvement of children in formal rather than informal activities (Valentine and McKendrick, 1997). The neighbourhood is no longer the most important playground for children; structured activities such as sports or music are increasingly emphasised.

These differences have spatial, temporal and economic impacts on men and women's everyday travel activities and use of transport. Social and cultural change surrounding these areas will therefore also have effects on everyday travel.

Data and method

The empirical analyses in this chapter are based on the Norwegian national personal travel survey (NPTS) from 1992 and 2005. The NPTS 1992 and 2005 consist of random samples of approximately 6000 and 17,500 people respectively, 13 years of age and above. The response rates were 68 percent and 50 percent, respectively.

The respondents were interviewed by telephone about their daily travel activities, household characteristics and personal information such as education, income, transport resources and employment. Both surveys shared the same main questions.

The personal travel survey is comprised of the following subjects: introduction, access to transport resources for the interviewed person and for the household, activities and travel the day before the interview (purpose, length, time use, transport mode, when and where the trip started and ended), long trips (100 km or longer during the previous month), employment/occupational status, the journey to work, education and employment of the spouse, information about the household (the number and age of children) and the interviewee.

Results – development from 1992 to 2005

Access to cars and fringe benefits

Access to transport means is in most cases the most important condition for travelling (even if about 20 percent of the daily trips are on foot). In 1992 a little more than 70 percent of the male population had continuous access to a car (with both driving licence and access to a car whenever wanted). Among women, 55 percent had the same level of access. In 2005 the figures were almost the same, 72 and 56 percent respectively. The result indicates no change. A closer look at the various age groups gives a more differentiated picture.

Table 13.2 Driving licence and always access to a car among men and women in different age groups (Norway, 1992 and 2005, percent)

Age groups	1992		2005	
	Men	Women	Men	Women
18-24 years	63	53	51	34
25-34 years	79	72	76	71
35-44 years	82	78	84	77
45-54 years	87	67	84	75
55-66 years	74	48	86	64
67-74 years	73	20	84	40
75 + years	54	9	70	15
All	70	55	72	56

While access to a car varied relatively little among the male age groups, the differences were much larger among women, especially in 1992. In this period women 45 years and older have increased their access to cars significantly, but for this group there is still potential for better access. The biggest difference is found in the oldest group, people 75 years and older. An interesting phenomenon is the decrease in access for the youngest group, both among men and women. This seems to be an international trend, at least in the Scandinavian countries (Hjorthol, 2002b, 3). The explanations have been related to economic conditions (Kranz and Vilhelmson, 1996, 19), increased environmental awareness (Ruud, 1999, 11), less interest in the car as a cult object (Sjöberg, 2000, 36; Nordbakke, 2002, 16), and more young people occupied with education and living in urban areas (Hjorthol, 1999, 55; Hjorthol, 2002b, 3).

There are still significant differences between the genders in relation to access to a car among people in the middle-aged groups (25–44 years), a difference that has not changed over the decade. What seems obvious is that the differences between the female age groups will be reduced when the current cohorts of younger and middle-aged women enter the older age categories. It is too early to say what will be the effect of young people's decreasing inclination to obtain a driving licence. The result of this analysis indicates that women have the potential to use a car to a greater extent than before, but that there remain significant differences between the genders.

In addition to having higher average incomes than women, men also have job-related benefits which make it more attractive for them than for women to use cars. In 1992, 17 percent of employed men had jobs which entailed car allowances in some form or another, whether company cars or expenses for the use of one's own car. Only 5 percent of women had equivalent benefits. In 2005, the figures were respectively 20 and 6 percent. In other words, employers paid for a rather large proportion of the journeys made by men.

Different trip purposes and trip length

In 1992 men made more trips per day than women, 3.4 versus 2.9. In 2005, both made 3.3 trips per day. The reason is first of all that more women had paid work and worked full time in 2005 than in 1992. Men, however, travelled much further than women. On average men travelled 42 km per day, while women travelled 29 km (2005). Women's 'area of action' is significantly smaller than men's. This suggests that women's movements are more local than men's.

Even if men and women take about the same number of trips, the purpose of the trips was still different, as is shown in Table 13.3 (the table does not include all reasons for travel).

Table 13.3 Number of trips per day for different purposes for men and women (Norway, 1992 and 2005)

Trip purpose	1992		2005	
	Men	Women	Men	Women
Paid work	0.71	0.40	0.71	0.57
Business	0.17	0.03	0.11	0.05
Shopping	0.78	0.84	0.87	0.97
Unpaid care work	0.23	0.27	0.32	0.36
Leisure	0.64	0.50	0.57	0.56
Social visit	0.47	0.50	0.40	0.46

Men made more work, business and leisure trips (escorting children to their activities is categorised as unpaid care work). Women made more trips related to shopping and unpaid care work, which mostly consists of escorting children to various activities. These differences did not diminish over the period. Women undertook more escort trips than in 1992, which indicates that women are assuming more of the family responsibilities for chauffeuring children. It appears that the increased access to a car for women in these age groups is related to such duties.

Household maintenance responsibilities were also still gendered. The cultural aspects of parenting or mothering discussed in part two were reflected in these type of trips. This increase is a sign of factors such as concern for children's safety and involvement in activities beyond the neighbourhood. But the increase might also be related to time pressure in daily life (Fyhri, 2002, 27) and the fact that driving is more convenient than other travel modes (Mackett, 2002). For children, increased car travel might result in poor health and obesity (Sleap and Warburton, 1993) or they may not 'learn' to know their neighbourhood or how to travel by themselves (Preiss, 1989, 102).

The use of transport means

The data show still significant differences between men and women in relation to use of transport means (see Table 13.4).

Table 13.4 Transport mode on everyday travel for men and women (1992 and 2005, percent)

Transport mode	**1992**		**2005**	
	Men	Women	Men	Women
On foot	18	25	18	23
Bicycle	7	5	5	4
Public transport	7	10	7	9
Car passenger	7	19	7	16
Car driver	59	41	62	47
Other	2	-	1	1
Sum	100	100	100	100

Table 13.4 also indicates that the change in transport mode from 1992 was rather modest. For both men and women there was a slight increase in driving a car, a little more for women than for men. Averages sometimes hide important information, however. When we look at the different age groups for both men and women, we see that there have been significant changes in some of them.

Except for the youngest group, 18–24 years, there has been an increase in car use for all (although for some middle-aged men, the status quo remained unchanged) (see Table 13.5). The increase has been largest among people 55 years and older, for both men and women.

Table 13.5 Car use as a driver on everyday travel for men and women (1992 and 2005, percent, differences between men and women in parenthesis)

Age groups	1992		2005	
	Men	Women	Men	Women
18-24 years	57	45 (-12)	44	31 (-15)
25-34 years	66	54 (-12)	64	53 (-11)
35-44 years	70	55 (-15)	73	65 (-8)
45-54 years	74	47 (-27)	74	59 (-15)
55-66 years	66	31 (-35)	74	48 (-24)
67-74 years	52	15 (-37)	73	33 (-40)
75 + years	41	4 (-37)	59	20 (-39)

Changes have been more modest for individuals in the age categories between 25 and 55 years, especially for men (some of them have also experienced a slight decrease). For men in these middle age groups, it seems as if car use had stabilised at a level of around 70 percent. For women in the same age categories, the increase was somewhat higher than for men, making the differences between the genders smaller. Among the oldest groups there has been a considerable increase in car use for both men and women, but the differences between the genders have increased (in percent points). Young men and women have both reduced their car use during this period, but women more than men. As shown in Table 13.2, this age group had less access to a car in 2005 than they had in 1992 and also, to a lesser degree, a driving licence.

Are women using cars for different purposes than men? Table 13.6 shows that men and women used their cars for the same purposes, but women less often than men. The differences between the genders are reduced for work-related trips and trips linked to caring for others (mostly children).

Table 13.6 Car use as a driver for various trip purposes for men and women (1992 and 2001, percent, differences between men and women in parenthesis)

Trip purpose	1992		2005	
	Men	Women	Men	Women
Work	70	53 (-17)	70	60 (-10)
Business	81	56 (-25)	73	54 (-19)
Shopping	63	42 (-21)	70	50 (-20)
Unpaid care work	92	72 (-21)	88	78 (-10)
Own leisure	38	21 (-17)	38	23 (-15)
Social visits	58	36 (-22)	57	39 (-18)

An interesting question is whether men and women who choose to use a car are in the same situation. I will present a multivariate analysis of escort trips to see if the conditions behind choosing car use were the same for men and women and if this changed during the survey period. Age, travel length, household income, education and quality of public transport will be taken into account.

In 1992 travel distance had no influence on the use of car for escort trips. The probability of using a car increased with age for men and decreased for women. Household income had a greater impact on car use for women than for men. While highly educated women used a car more than women with less education, the opposite was the situation among men. Men used a car even if public transport was fairly good, while women's use of a car was significantly lower in areas with high quality public transport.

In 2005 the picture was a bit different. Among young women there was still more car use than among the elderly, but this was also true for men. The importance of household income was reduced for both men and women in determining car use, while education still had a significant impact on women's car use. Car use in 2005 for both women and men was lower in areas with good public transport than in areas with poor public transport.

Table 13.7 Use of car on chauffeuring trips (unpaid care work) for men and women aged 18 or older (1992 and 2001, logistic regression)

Variables in the analysis	1992 B		2005 B	
	Men	Women	Men	Women
Travel length	-.003	.009	.004	-.009*
Age	.038*	-.026**	-.014**	-.012*
Household income	.011	.193***	.126*	.076*
Education[1]				
Low	1.044	-1.073***	.089	-.213
Middle	1.212**	-.170	.118	.196
Quality of public transport[2]				
Very good and good	.099	-1.246***	-.934***	-.752***
Fairly good	.760	-.436	-1.118***	-.093
Constant	.210	1.882***	1.595***	1.209***
N	568	725	2473	2652

*** p<.000. ** p<.001. * p<.05

Notes: [1] higher education, university level; [2] Bad or very bad quality.

The effects of the different variables on car use for escort journeys seems to have become more alike for men and women during the period. This fact indicates that when women gain better access to private transport resources (a car) they often use this access to benefit others, as in chauffeuring children and the elderly to activities.

The work trip

In 1992, 22 percent of women combined their trips to work with grocery shopping, while 9 percent escorted children to and from school or kindergarten. The corresponding figures for men were 15 and 5 percent. In 2005, 17 percent of the women shopped in relation to the journey to work and 8 percent escorted children. Among men the figures were respectively 11 and 5 percent. The differences between the genders persisted, but it was smaller than in 1992.

As Table 13.6 indicates, the differences in car use on the journey to work have changed little. Women increased their car use to some extent during the period. To examine if the effects of different variables on car use have changed during the period, multiple regressions were employed on data for both men and women.

In addition to age, education, working hours, commuting distance and individual income, the variables in the analyses were related to family situation, the residential context and the quality of the public transport supply (see Table 13.8).

Table 13.8 Use of car as a driver on work trips for men and women with a driving licence aged 18 years or older (1992 and 2001, logistic regression)

Variables in the analysis	1992 B		2005 B	
	Men	Women	Men	Women
Travel length	.019***	.038***	-.001	.005**
Age	.002	-.006	.007*	.001
Own income (gross)	.003	.026	.1221***	.255***
Working hours per week	.021 **	-.016	.019***	-.012**
Education[1]				
Low	.139	-.412*	.748***	-.030
Middle	.328*	-.416*	.592***	-.002
Quality of public transport[2]				
Very good and good	.031	-.464*	-.359***	-.674***
Fairly good	-.046	-.109	-.263**	-.468***
Residential area[3]				
Oslo. Bergen. Trondheim	-.553**	-.739***	-.593***	-.603***
Surr. Area of Oslo/Be/Tr	-.421*	-.310	-.055	-.044
Type of family[4]				
Single	.417**	-.387*	-.333**	-.144
Couples without children	.248	-.293	-.524***	-.498***
Single with children	.718	.188	.797**	.800***
Constant	-.387	1.664***	-.337	.508*
N	1563	971	4975	3668

*** p<.000. ** p<.001. * p<.05

Notes: [1] higher education, university level; [2] Bad or very bad quality; [3] Rest of the country; [4] Couple with children

Conditions related to family situation, travel length, the area where people live and the quality of public transport were the most important factors governing car use for work trips for women in 1992. Single women and married/cohabitating women without children used a car less than women with children. Women in the largest cities and those who had access to good public transport used a car less than other women. Travel length also led to an increase in car use. In 1992, women with a university-level education used a car for work trips more often than women with less education. But for men in 1992, travel length, working hours and place of abode were important variables for car use. Like women, men increased their car use with travel distance. They also used a car less in the largest cities. But the quality of public transport did not affect car use when commuting. Unlike women, single men used a car more than men with children, which indicated that men were less responsible for taking children to day care and school than women in 1992.

In 2005 the picture was somewhat changed. For women, travel length, income, working hours, quality of public transport, place of abode and family situation affected car use for commuting. Car use increased with travel distance and income, but declined with working hours. As was true in 1992, women living in the largest cities and in areas with good quality public transport used a car less than women living in other areas. Women with children, both single and married, used a car more than women without children. In terms of women's car use, children had an even greater impact on car use in 2005 than in 1992.

There have also been changes during this period in the variables that affect men's car use in commuting. Educational levels did affect men's car use in 2005, where men with the lowest and middle levels of education used a car more frequently for their work trips than men with higher levels of education. The reason may have been relocation of the workplace from cities to more peripheral areas. The quality of public transport also affected men's car use in 2005. But the effect of this variable was greater for women than for men. The most interesting change from 1992 to 2005 was related to men's family situation. As was true for women, but unlike the 1992 results, men with children, and especially single men with children (very few) used a car for work trips more often than men in other family situations. This result indicates that family situation also affected men's commuting practices.

To summarise, this analysis indicates changes in both men's and women's car use on work trips. But differences remain. Travel length and the quality of public transport had more of an impact on car use for women than for men, but family situation seemed to have the same effect for both genders. This result suggests that men's car use was more independent of structural variables than women's car use, but that men were more involved in everyday responsibilities for children in 2005 than in 1992.

Discussion

Today's society demands mobility, and the emphasis on the car as the dominant means of transport has both environmental and social consequences. As Doreen Massey writes: '…every time someone uses a car, and thereby increases their own mobility, they reduce both the social rationale and the financial viability of the public transport system – and thereby also potentially reduce the mobility of those who rely on that system' (Massey, 1994, 150).

The analyses in this chapter reveal that women and men's travel patterns reflect their different social positions. Travel and use of transport modes resulted from the individual's position in the labour market, the division of unpaid work in the household and access to resources. Despite achievements in increasing equality between the genders, work and daily life situations still differ between men and women. The differences in average income and access to work-related fringe benefits were as large in 2005 as in 1992. Even while most Norwegian women have paid work and a greater percentage of them worked full-time in 2005 than in 1992, women to a greater extent than men still have the main responsibility for the 'management' of the household, even in families without small children, a fact that is amply demonstrated by both time use studies and the elucidation of women's reasons for daily travel and mobility. As long as the distribution of tasks between genders within a household are maintained, the differences will also be reflected in travel patterns.

The results from time use surveys showed that women spent less time on household work while men spent more time on their children. This indicates that there is some change in the distribution of household work, but that the process is slow, and for the time being has had little effect on daily mobility patterns. These very small time use changes are not unique to Norway; they are found in several western countries.

The study of men and women's daily travel patterns can be seen as a 'barometer' of the state of equality between men and women in society. Even while the number of trips have increased for women during the period of analysis and have now reached the same level as for men, women have a much more limited 'action space' than men. This might restrict their labour market choices, which thus reinforces differences. The analysis also shows that access to a car is still lower for women in most age groups than for men. Little has changed during the 1990s in Norway. The increase in access to a car for some groups of women has resulted in more family chauffeuring. The driving of children to kindergarten, school and leisure activities has increased over the period, and has become a task for women more than men. Time pressure in families with children is said to be an important reason for increased car use (Hjorthol et al., 2006, 42). While women make more trips *for others*, men make more trips *for themselves.* In families with only one car, the husband is the primary user (Hjorthol, 2000). However, the multivariate analysis in this chapter of the changes in car use for commuting indicated that family situation had a greater impact on men than women.

The analysis revealed a slight increase in car use among women, but there is still a greater potential for growth compared to men's use. If a car is necessary for managing everyday life and provides access to the labour market, it is hard to believe that women do not want a car for themselves. The increase in two-car households was significant. For the time being the difference between men and women's *use* of a car is greater than their *access.* One explanation may be related to habit. The more one is accustomed to using a car, the more frequently it is used for all purposes. Norwegian women have probably not become as accustomed to car use as men have. Traditionally, interest in cars has been much greater among boys and men than among girls and women. As indicated in Table 13.2, young women did not give priority to obtaining a driving licence and a car to the same degree as young men did. It is difficult to tell if this is a trend that will persist. But it is clear that women more than men see the car as merely a practical device, which suggests less ownership of cars among women than men.

At the same time, surveys about environmental engagement and attitudes towards environmental problems suggest that women more than men are concerned about these issues (Polk, 1998, 123–5; Matthies et al., 2002). Surveys of the relationship between environmental attitudes and action and studies of collective versus individual attitudes indicate that women to a greater degree than men choose to act in an environmentally friendly manner (Davidson and Freudenberg, 1996; Garvill et al. 1994:20). Environmental arguments against car use continue to increase. The question is, however, if changes in environmental attitudes also will result in changes in car use. When women express more environmentally friendly attitudes than men, is it because they are more engaged in environmental questions than men, or is it because this attitude is more in accordance with their ability to act? Will their attitude become more like men's if/when they gain the same access to a car as men? And what about men – will a more restrictive car policy influence their environmental attitudes? These questions should be of great interest for transport policy.

The analysis in this chapter demonstrates the importance of taking gender aspects into consideration in transport planning and policy. Even while mobility patterns for the genders are converging, the differences are still significant. The trend, however, is for development that is more 'masculine' than 'feminine', and as such is less environmentally friendly. A transport (and land use) policy emphasising collective solutions might benefit both gender equality and the environment.

References

Andrews, H.F. (1978), 'Journey to work considerations in the labour force participation of married women', *Regional Studies* 12, 11–20.

Cristaldi, F. (2005), 'Commuting and Gender in Italy: A Methodological Issue', *The Professional Geographer* 55, 268–84.

Dasgupta, M. (1983), *Travel-to-Work Characteristics of Different Labour Force Groups: A survey in Manchester* (TRRL Laboratory Report Report 1069. Crowthorne, Berkshire).

Davidson, D.B. and Freudenburg, W.R. (1996), 'Gender and environmental risk concerns. A review and analysis of available research', *Environment and behaviour* 28, 302–39.

Dobbs, L. (2005), 'Wedded to the car: women, employment and the importance of private transport', *Transport Policy* 12, 266–78.

Ericksen, J. (1977), 'An analysis of the journey to work for women', *Social problems* 24, 428–35.

Faganini, J. (1987), *Daily commuting time. The stakes for working mothers in France* (Washington DC: National Research Council. Transportation Research Record 1135).

Fyhri, A. (2002), *Barns reiser til skolen. En spørreundersøkelse om reisevaner og trafikksikkerhet på skoleveien. (A survey of travel behaviour and traffic safety on the school road)* (Oslo: Institute of Transport Economics, report 616/2002).

Garvill, J., Laitila, T. and Brydsten, M. (1994), *Livsvärden och val av transportmedel* (*Life values and choice of transport means)* (Umeå Universitet: Transportforskningsenheten).

Gera, S. and Kuhn, P. (1978), *Occupation, locational patterns and the journey-to-work* (Ottawa: Economic Council of Canada. Discussion paper no. 121).

Guiliano, G. (1979), 'Public transportation and the travel needs of women', *Traffic Quarterly* 33, (4), 607–16.

Hagman, O. (1995), *Bilen och naturen i svensk bilreklam. En studie av bilsamhällets mytologi (Car and nature in Swedish car advertising)* (Gothenburg: The University of Gothenburg, Socialantropologiska institutionen. Socialantropologiska skrifter).

Hanson, S. and Hanson, P. (1980), 'Gender and urban activity patterns in Uppsala, Sweden', *Geographical Review* 70, 291–9.

Hanson, S. and Johnston, I. (1985), 'Gender differences in work-trip length. Explanations and implications', *Urban Geography* 6, (3), 193–219.

Hanson, S. and Pratt, G. (1988), 'Spatial dimensions of the gender division of labor in a local labor market', *Urban Geography* 9, (2), 180–202.

Hanson, S. and Pratt, G. (1995), *Gender, work, and space* (London and New York: Routledge).

Harvey, D. (1989), *The condition of postmodernity. An enquiry into the origins of cultural change* (Oxford UK/Cambridge, Mass, USA: Basil Blackwell).

Hjorthol, R. (1998), *Hverdagslivets reiser. En analyse av kvinners og menns daglige reiser i Oslo (Travel activities in everyday life. An analysis of women and men's daily travel activities in Oslo* (University of Oslo: Oslo, Institute of Transport Economics, TØI report 391).

Hjorthol, R. (1999), *Daglige reiser på 90-tallet. Analyser av de norske reisevaneundersøkelsene fra 1991/92 og 1998/98 (Daily travelling in the 90s. Analyses of the Norwegian personal travel surveys 1991/92 and 1997/98 (*Oslo: Institute of Transport Economics, TØI report 436/1999).

Hjorthol, R. (2000), 'Same city – different options. An analysis of the work trips of married couples in the metropolitan area of Oslo', *Journal of Transport Geography* 8, 213–20.

Hjorthol, R. (2002a), 'Cultural aspects of the car and public transport. Men's and women's perception', *Raumplanung* 102, 144–9.

Hjorthol, R. (2002b), *Ungdom og skolereiser. Den nasjonale reisevaneundersøkelsen 2001 (Young people and school trips in Norway. The national personal travel survey 2001,* (Oslo: Institute of Transport Economics, TØI report 597/2002).

Hjorthol, R., Hovland Jakobsen, R. and Ling, R. (2006), *På farten – i bilen – med mobilen. En studie av kommunikasjon og mobiliet i barnefamiliers dagligliv (On the move – in the car – with the mobile phone: A study of communication and mobility in the daily life of families with children)* (Oslo: Institute of Transport Economics, TØI-report 820/2006).

Hjorthol, R. and Sagberg, F. (1998), *Changes in elderly persons' modes of travel* (Oslo: Institute of Transport Economics, TØI working report 1105/998).

Johnston-Anumonwo, I. (1992), 'The influence of household type on gender differences in work trip distance', *Professional Geographer* 44 (2), 161–9.

Kranz, L.-G. and Vilhelmson, B. (1996), *Förändringar av den dagliga rörligheten i Sverige 1978-1994* (Göteborg: Göteborgs Universitet, Kulturgeografiska institutionen).

Kwan, M.-P. (2000), 'Gender differences in space-time constraints', *Area* 32, 145–56.

Lamvik, G.M. (1994), *Et eventyr på hjul. Et antropologisk blikk på amcar-kulturen (An adventure on wheels. An anthropological view of the Amcar culture)* (Report nr 21. Trondheim: The University in Trondheim, Centre for technology and society).

Lash, S. and Urry, J. (1994), *Economies of signs and space* (London: Sage).

Law, R. (1999), 'Beyond 'women and transport': towards new geographies of gender and daily mobility', *Progress in Human Geography* 23 (4), 657–88.

Mackett, R.L. (2002), 'Increasing car dependency of children. Should we be worried'? *Municipal Engineer* 151 (1), 29–38.

Madden, J.F. (1981), 'Why women work closer to home', *Urban Studies* 18, 181–94.

Madden, J.F. and White, M.J. (1979), *Women's work trips. An empirical and theoretical overview in women's travel issues. Research needs and priorities* (Washington DC: US Dept of Transportation).

Manning, I. (1982), *The journey to work* (Sydney: Allen & Unwin).

Massey, D. (1994), *Space, place and gender* (Cambridge: Polity Press).

Matthies, E., Kuhn, S.D., Klöckner, C.A. (2002), 'Travel mode choice of women: The result of limitation, ecological norm, or weak habit?' *Environment and Behavior*, 34 (2), 163–177.

Nordbakke, S. (2002), *Førerkort og bilbruk blant ungdom på 90-tallet. Tegn på endringer i ungdoms reisevaner?* (Oslo: Institute of Transport Economics. TØI report 564/2002).

OECD (2001), *Aging and transport. mobility needs and safety issues* (Paris: OECD).

OECD/ECMT (2000), *Transport and the aging of the population* (Round Table 112, Paris: European Conference of Ministers of Transport, OECD).

Polk, M. (1998), *Gendered mobility. A study of women's and men's relations to automobility in Sweden* (Gothenburg: Gothenburg University, PhD dissertation, Humanekologiska skrifter no 17).

Preiss, D. (1989), 'Frauen – Kinder – Auto – Träume', in: Beckmans et al., *Welche Freiheit brauchen wie? Zur Psychologie der AutoMobilen Gesellschaft* (Berlin: Vas in der Elefanten Press).

Rosenbloom, S. (1987), 'The impact of growing children on their parents' travel behaviour: a comparative analysis', *Transportation Research Record* 1135, 17–25.

Rosenbloom, S. (2002), 'Asking the right question – Understanding the travel needs of older women who do not drive', *Transportation Research Record* (1818), 78–82.

Rosengren, A. (1993), *Some notes of the male motoring world in a Swedish community.* Paper presented at the workshop: 'The car and its environments. The past, present and future of the motorcar in Europe', Trondheim, May 6–8.

Rutherford, B. and Wekerle, G. (1988), 'Captive rider, captive labor: spatial constraint on women's employment', *Urban Geography*, 173–93.

Ruud, A. (1999), *Bilismen er skadelig for miljøet – men spiller jeg noen rolle?* (Oslo: Institute of Transport Economics, TØI report 424/1999).

Sirmans, C.F. (1977), 'Determinants of journey to work flows. Some empirical evidence', *The Annals of Regional Science* 11 (1), 98–108.

Sjöberg, A. (2000), *Färdsätt. En etnologisk studie om resvanor, miljömedvetande och livsstilar bland unga i Göteborg* (Göteborg: Etnologiska föreningen i Västsverige).

Sleap, M. and Warburton, P. (1993), 'Are primary school children gaining heart health benefits from their journey to school?' *Child: Care, Health and Development*, 19, 99–108.

Urry, J. (2000), Sociology beyond Societies. Mobilities for the Twenty-first Century (London and New York: Routledge).

Urry, J. (2003), 'Social networks, travel and talk', *British Journal of Sociology* 54 (2), 155–75.

Vaage, O.F. (2002), *Til alle døgnets tider. Tidsbruk 1971-2000 (Time use 1971-2000)* (Oslo-Kongsvinger: Statistics Norway, Statistical analyses 52).

Valentine, G. and McKendrick, J. (1997), 'Children's outdoor play; exploring parental concerns about children's safety and the changing nature of childhood', *Geoforum* 28, 219–35.

Vandersmissen, M.-H., Villeneuve, P. and Thériault, M. (2003), 'Analysing changes in urban form and commuting time', *The Professional Geographer* 55, 446–63.

Villeneuve, P. and Rose, D. (1988), 'Gender and the separation of employment from home in metropolitan Montreal: 1971–1981', *Urban Geography* 9 (2), 155–79.

Wachs, M. (1992), 'Men, women, and urban travel. The persistence of separate spheres', in M. Wachs and M. Crawford (eds), *The Car and the City. The Automobile, the Built Environment and Daily Urban Life* (Ann Arbor, MI: University of Michigan Press).

PART 3
Seeking Grounds for Future Policies

Chapter 14

Gender and the Social Usage of Mobile Technologies: From Information Society Policies to Everyday Practices

Tommi Inkinen

Introduction

Some ten to fifteen years ago, social theorists interested in technology claimed that the Internet, mobile phones and other information and communication technologies (ICTs) would revolutionise social relationships. Subsequently, several studies (May, 2002; Hannemyr, 2003; Thulin, 2004; Evans and Brooks, 2005) clearly showed that the changes in everyday life have not been so dramatic. However, this is not to say that there have not been changes at all, or to underestimate them. Mobile communications have extended the concept of mobility from physical movement to social interactions, and have also changed several practices in daily life. The observation of these changes is not always straightforward and has also challenged researchers to develop new methodologies in order to understand the dynamics of these changes.

Since the beginning of the 1990s, Finland has made its mark as one of the leading mobile phone producers in the world, mainly due to the Finnish company Nokia's global success. As a technology, the mobile phone is closely associated with the concept of the 'information society' (IS). For example, Kasesniemi (2003, 14) states that mobile phones can be seen as 'a passageway' to the IS. The information society is a macro-level term that is commonly associated with business, the economy, information and data transfer, and technological development – issues that do not reflect aspects of gender, race, ethnicity or other social and cultural dimensions that are the essentials of the societal fabric. The information society is a concept widely used to describe the society in which we live. Therefore, it needs to be viewed from various perspectives, particularly accounting for the diversity of social conditions.

Most of the impacts that mobile technologies have are spatial. A clear indication of this is the contextualisation of mobile calling. The current socio-spatial location of the answerer is one of the most popular opening phrases: 'Where are you and are you able to talk right now?' This phrase intertwines the social (talk) and spatial (where) with the temporal (now), because the location in time-space has social borders with 'etiquettes'. Some locations are more sensitive than others. Knowing the identity of the caller is also important: people respond depending on who is calling. Both spatial and social contexts play a role in human behaviours and interactions.

Thus there are two rather different fields to explore. First, it is necessary to consider the social realm as a setting for everyday life with respect to national legislation, cultural conduct and societal practices. This realm is also geographical – the lived location within the national context. Second, we have individual practices – the subjective actions: 'What do I do and why do I do it that way?' Personal histories, visible and invisible power relations, structural codes and personal choices form the ever-changing mix of choices and their outcomes, resulting in aggregates of organisations, collective arrangements and their inter- and intra-relations. Thus, the micro-level collectively transfigures into the macro-level in time and space.

In this chapter, I will address the connections between three key elements: mobile (technology) practices in everyday life, the role of gender in these practices and information society policy. These elements may seem to be quite distant from each other – but they are not. As a matter of fact, this triad describes the descent from policy (macro) to individual everyday practices (micro). I will discuss everyday practices of mobile technology use with reference to the latest available information society strategy paper of the Finnish government.

Finland makes a good platform for case data because it has been regarded as one of the leading countries in the development of an information and knowledge society (for example OECD, 2005, 11). Finland is also regarded as one of the least corrupt countries (Transparency International, 2006, 154–6). Gender equality is also one of the top priorities of the national ethos. My aim is to present examples through which it might be possible to further develop policy options that would give gender issues more prominence in policy guidance for the information society.

To meet these objectives I will first review some of the literature on gender and technology use. I then provide a background for understanding Finnish information society policy. Here I examine how the social dimension has been taken into account and how gender issues are present in the latest policy document. Third, I provide empirical survey results of gender differences in everyday practices of mobile technology use. I use the selected statements to analyse technological mobility and contrast the results with the Finnish information society strategy. The chapter concludes by addressing issues relevant to future research and policy creation.

Mobility, gender and difference of networking

The conceptual framework of this chapter is based on a discussion regarding two key terms. These are information and communication technology based 'mobility' and 'networking'. A crude description of the two would probably be the separation into technology (devices) and content (digital formats) that carry and deliver the meaning (significance or experience) of the communication. The first essential concept is mobility. Mobility is a broad concept that refers to several aspects of human life (other chapters in this book; also Kellerman, 2006; Larsen et al., 2006). These include aspects of travel of people for work, leisure, migration and refuge. Mobility includes also movement of physical and immaterial goods creating a need for transportation and logistical systems. It also involves communicative and digital dimensions such as virtual travelling on the Internet, experiences of imaginative travelling elsewhere through television broadcasts and communication via personal mobile technologies such as mobile phones. In this chapter I use the term 'mobility' in a restricted sense to refer to communicative practices and experiences mediated through information and communication technologies. The empirical data concerns the use of the mobile phone, which is one of the predominant technologies that constitute the essence of 'mobile communication' by bringing together text, voice and the Internet (Inkinen, 2003, 30).

Considering current technologies used in mobile communication there are several technical terms and descriptions: Global System for Mobiletelephones (GSM) is the current standard of mobile devices in Europe; it is also called 'second generation'. The 'third generation' (3G) networks are currently being developed and implemented to provide considerably faster data-transaction capabilities to mobile devices. Associated terms that are the most relevant for ordinary people, in the role of 'consumers' of mobile services, are Short Messaging Service (SMS) and Multimedia Messaging Service (MMS). Several customer services combining visual and audio components with user interaction have been created in step with the growing data processing capabilities of mobile devices.

For the purposes of mobile phone use, there are several ways to conceptualise mobility in terms of technology and communication. For example, mobility can be divided into two categories as proposed by Kasesniemi (2003, 36–7). These two categories are 'semimobility' and 'full mobility'. The definition of these categories is based on the possibility of concentrating fully on communication with reference to physical movement:

> ...a person driving a truck is an example of the fully mobile communicator who is engaged in physical movement and by consequence cannot give their undivided attention to communication. A mobile communicator located in a train, hotel room or in someone else's office functions in a state of semimobility. The environment may be entirely stable, with the use of wireless communication motivated solely by the fact of being away from familiar surroundings; other communication channels dependent on equipment accessible at home or in the office may not be available. Use is thus not induced by mobility but by the lack of other communication channels. (Kasesniemi, 2003, 37)

The quote demonstrates the idea of thinking about 'mobility' in terms of physical movement, the spatial situation and opportunity to use technologies in a changing context. The division into two categories is, however, not without problems because the definition relies mainly on the physical attributes of the mobile technologies in question. A mobile phone is light and small, and thus portable. This portability leads directly to the combination of the user and the technology with changing spatial contexts dependent on movement. In technical terms, it is necessary to broaden the 'mobility' concept of this chapter to also involve portable laptop computers and palm devices. These are not as common as standard mobile phones but computer technologies, wireless data transaction infrastructures and deliverable contents together form the platform for mobile communications.

Gender has for a long time been one of the key social dimensions in the analysis of social technologies (Haraway, 1991; Wajcman, 1991; Turkle, 1995; Spender, 1995). In a more recent study, the gendered construction of mobile technologies has been analysed by using interview data from designers and marketers in the ICT industry (Mörtberg, 2003, 158). Mörtberg's analysis is thus based on the assumptions of the technology development experts. Her analysis of expert opinions regarding mobile technologies and their relation to gender is sound supportive material for the results obtained in this survey. She uses the concepts, or 'figurations' as she puts it, of feminist technology research, which are nomadic subjects and cyborgs derived from the thinking of Haraway (1991). The central conclusion highlights the importance of gender in the process of technology creation:

> The networks of sociotechnological relations are consequences of today's science and technology dependent society. The networks are created and sustained by high tech systems which can facilitate the reconfigurations of gender, ethnicity and class [...] The stories created in the designers' visions appear very early in the design of an IT-system, services or artefact, usually before the present domain or situation is analysed [...] The women and men interviewed seemed to reconstruct technology as masculine in a general level. However, this was dominating since on the more concrete level, one's work tasks, the nomadic subject contributed to thinking beyond and through existing categories or images of gender and technology. The images and understanding of gender were not homogenous categories. In the interviews, gender differences were reconstructed but gender dichotomies were also rejected. Thus, categories are not fixed but are (re)created and (re)constructed in ongoing processes. There is no transparency between reality and categories since something always exists outside stabilization. (Mörtberg, 2003, 166–7)

A key point in the quote is the criticism of fixed categories for social entities. Interestingly, the developers and marketers of technology evaluated mobile technology as being masculine in its essence mainly due to the gender bias in the workforce of high-tech industries. Technology is commonly created in male-dominated production facilities. This is due to the still small proportion of women in technical disciplines. As an example, in Finland in 2005, only 25% of students gaining a master's degree in technology were women (Statistics Finland, 2006a).

An important issue is the purpose of the mobile communication – the communication content. The key element of the survey is to obtain an idea of the practices of social networking through mobile devices. To refer to 'social networking' immediately raises a number of problems of interpretation. Social networking is a holistic term. I use it mainly to refer to the contents of mobile communication between individuals.

For ordinary individuals, the predominant method of mobile communication is voice calling. The other major means of mobile communication are based on textual typing (SMS, email) or other digital transactions (MMS and software uploading or downloading). In technical terms, voice is also digital data transaction. However, the voice call is essentially an instant form of communication with two or more individuals (group calls). In the case of answering machines and 'missing a call', the voice communication can be considered to be 'delayed'. The delay is essentially typical in typing-based forms of communication. Typed communication is also always textual and thus can be studied by means of textual analysis (for example Kasesniemi, 2003, 15).

The dynamics of instant and delayed with voice (audible) and textual (silent) communication result in differentiated usage patterns, because the mobile communication occurs in changing socio-spatial contexts. Thus the social location and condition of the origin of communication underlie the observed outcome of human-to-human communication. Mobile communication can also be based on human-machine or machine-machine interactions. Internet browsing and software downloading are good examples of the former, while automated wireless Bluetooth file-sharing is a good example of the latter: mobile devices can be 'taught' to automatically share picture or audio files, for instance, if the authenticated units (such as phones and lap-tops) are within range of each other.

The empirical part of this chapter concerns the differences between genders in mobile technology use. Gender differences occur in the methods of communication (voice, data, SMS), as will be shown in the empirical section. Thus the gender differences in social communication are of key interest. Lieber and Sandefur (2002), for example, have studied gender differences in the exchange of social support within several groups such as friends, neighbours and co-workers. After an extensive statistical analysis they concluded that women tend to share more social support than men. Their results confirm earlier research concerning similar topics by Wellman and Wortley (1990) and Barbee et al. (1993). However, they also state that there are several explanatory factors behind the process of social support exchange. These include marital status, socio-economic conditions and general involvement in social exchanges (Lieber and Sandefur, 2002, 388).

Similar results and indications have been obtained from various studies concerning social interactions, power relations and telephone and computer usage (for example DeSolaPool, 1981; Cox and Leonard, 1993; Grint and Gill, 1995; Hey, 1997; Dimmick et al., 2000; Inkinen, 2003; Morgan et al., 2004). Thulin (2004) studied young people and gender issues in her doctoral thesis, analysing a group of young people between the ages of 18 to 20. Her empirical results clearly show the tendency towards differentiation in communication frequencies and contents according to gender. According to her study, women (20-year-olds) use the Internet approximately 25 minutes per day whereas men use it for over 80 minutes. The usage gap tends to increase with time. The situation changes radically if we consider mobile telephony. Women have almost twice as many social contacts per day than men. If we consider means of communication, SMS messaging is far more popular among young women than men. There is also a significant difference in voice communication among 18-year-olds (girls talk more than boys) but in the group of 20-year-olds the difference diminishes. (Thulin, 2004, 74–84).

Based on the reviewed literature, I will use the following thematic in order to obtain a picture of gender differences. My approach is based on the presented aspects of technological mobility. I will use an approach similar to that used by Thulin in her dissertation in which she conceptualises the usage of ICTs in everyday life as a two-sided process. I will use a modified conceptualisation of Thulin's framework to classify my empirical material (Figure 14.1).

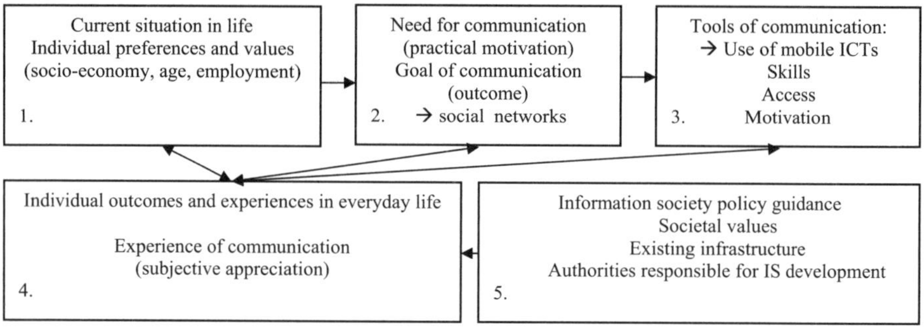

Figure 14.1 A conceptualisation of key dimensions of ICT use and everyday life
Source: modified from Thulin 2004: 43.

There are five essential components in the framework. The first four deal with individuals. The fifth box symbolises the general policy guidance of governments having an effect on everyday practices. Policy guidance is also dependent on the individual practices and thus should represent at least to some extent the existing social conditions in society. Unfortunately, this is not always the case.

Notes on gender and information society policy guidance in Finland

I have discussed some dimensions of mobility, social communication and gender. These issues and related processes take place in everyday life – at the (individual) micro-level. The second task of the chapter is to look at the central policy guidance documents for the development of the Finnish information society – the macro-level. Finland's information society has been developed principally through the implementation of government-produced strategies. Strategic thinking spread in Finland in the 1980s and a veritable golden age of strategies began in the early 1990s. The active creation of information society strategies did not start until the latter part of the 1990s (Chatrie and Wright 2000, 13).

Guidance for the development of Finland's information society strategy is overseen by the 'Information Society Council', an elite group composed mainly of politicians, representatives of IT businesses and to a lesser extent, researchers. Finland has not had a uniform national information society strategy since 1998, when the 'SITRA strategy' was published. SITRA is the Finnish National Fund for Research and Development, which operates under the supervision of the Finnish Parliament. The outlines presented in the SITRA strategy relied heavily on over-optimistic assumptions for the 'digital revolution' (May 2002, 1–3) and societal implications of information processing technologies.

After the SITRA strategy, the most important public sector action taken was when four 'government policy programmes' were launched in 2003; one of these has been the 'Information Society Policy Programme', which is overseen by the Prime Minister. The other three are policy programmes for employment (led by the Minister of Labour), entrepreneurship (Minister of Trade and Industry) and civil participation (Minister of Justice). The Information Society Policy Programme has defined its goals with reference to social and human-centred issues. The policy programme defines its mission as follows:

> The aim of the Government Information Society Programme is to boost competitiveness and productivity, to promote social and regional equality and to improve citizens' well-being and quality of life through effective utilisation of information and communications technologies. The Information Society Policy Programme aims to maintain Finland's status as one of the leading producers and users of information and communications technology […] The Programme is based on close cooperation between public and private sectors […] The Information Society Council and its sections contribute in a significant manner to this cooperation. (Government Policy Programmes, 2006, 1)

Table 14.1 Frequencies of gender and social welfare related terms with selected benchmark terms in Finnish information society strategy for the years 2007–2015

Term	Hits	Remarks
Gender	0	There were no hits for gender (sukupuoli in Finnish)
Women	2	'Women and men' were used as subcategories for age groups '64 to 74' and 'above 74'
Gender equality	0	
Equality	8	Of which 2 hits were for regional equality and 6 hits for social equality
Safety	7	Of which 1 hit concerned traffic security and 6 information security on the internet
Welfare	7	Of which 3 hits concerned individuals and 3 e-welfare services
Competitiveness	34	These terms are intended to show the most frequent terminology in the strategy
Business	38	
Service	220	
Content	17	
Technology	44	
Information society	231	

Currently, the Information Society Council has finalised its 2007–2015 strategy. The new strategy was published in September 2006; I will focus on this document. I used word processing software in order to find out the frequency of key terms[1] used. The terms used and the number of hits are presented in Table 14.1.

The results provide only a simple picture of the essence of the strategy paper. However, even this basic word search provides an indication that gender or even social welfare issues are *not* a 'hot' issue in the strategy. The result seems to contradict the general goals of the Information Society Policy Programme. Actually, it is very surprising that there was no mention of gender issues. In general, the term 'social equality' was used six times, but it was listed as a 'common goal' for society, among other noble goals such as sustainable development, strong economic growth and well-being. The two hits found for the term 'women' were in a table showing the amount of Internet use among age groups. On two occasions the oldest (60–74 and above 74 years) age groups were divided into gender categories and the gender issue was presented as a statistical figure with no further discussion.

If the strategy paper is read thematically, there are several points highlighting the importance of citizenship in the development of an information society. However, the meaning of citizenship has not been defined and there are actually no clear indications of how citizens should then be taken into account in the strategy. The strategy takes a rather simple approach and declares that a citizen-based information society will develop when more broadband connections and mobile services are developed. The citizen-oriented information society is thus being built mainly through the IT industry and its development will result in the production and selling of more 'mobility' to the masses. The role of the public sector is to produce e-services and create legislation concerning the information security. This type of strategy language is clearly visible just by looking at the frequency of hits for the six lowest key terms in Table 14.1.

The only social category taken into separate consideration in the strategy is age. Otherwise 'social' is a unified term including all human-related aspects. Elderly people are discussed and outlined as a specific target group. Other discussions related to social divisions or categories are not present. This raises the question of whether or not there should even be such discussions. However, empirical studies and theoretical considerations have shown that there are existing gender differences in technology use (Haynes et al., 2005; Cohoon McGrath and Aspray, 2006) and in the way that gender is presented in the media regarding that use (Jauhiainen, 2007).

It is also worth discussing the process of strategy creation. The strategy was mainly based on four expert meetings, so-called 'future workshops' for the information society. The executive group leading the strategy creation included 15 men and 5 women, with an evident gender imbalance. It is also interesting that the female representatives were all from public organisations or labour unions while the men were from the private sector. In this sense the gendered construction of the information society seems evident on the strategy level: the executive level of strategy creation was dominated by men who also represent the 'hard values' of economy and business. This interpretation is similar to other studies on policy creation and gender (Abbot and Sapsford 1988, 178–9; Mörtberg et al., 2004: 166–7; Archibald et al., 2005).

1 The strategy paper is currently available only in Finnish and the terms used are translations.

Gender and the use of mobile technologies in everyday life

A brief overview of preliminaries: data and approach

I used empirical survey data that was collected in 2004 from the Tampere region, one of the technological development cores of Finland (OECD, 2005, 35). There are three units of higher education in Tampere including the Tampere University of Technology, which is the second largest technical university in Finland. Extensive background descriptions of the case area, its socio-economic features and geography can be found on the Internet for the city of Tampere at <www.tampere.fi/english> and for Tampere region at <www.pirkanmaa.fi/english>.

The data is available from the Finnish Social Science data archive (FSD, 2007). Only one questionnaire was sent per household, and the target group was the working-age population. The data included a total of 2,000 questionnaires out of which 52.6% were returned after a second collection round. The percentage of female respondents was 55.9; this can be regarded as 'a normal bias' in survey studies considering that the percentage of women in the target population is 52.3 (Statistics Finland, 2006b). If compared to the data used by Lieber and Sandefur (2002, 374) the bias is rather similar (approximately 54%).

The content structuring (questions asked) was not originally built around the framework used in this chapter. Therefore I use selected statements that fit the thematic of Figure 14.1 and connect them to it. The statements concern boxes 2 (need for communication), 3 (tools of communication) and 4 (experience of communication). I present the results of the statements that have the largest gaps between women and men.

This type of analysis has some limitations. The sample is limited to a specific area in Finland and it cannot be automatically generalised for the whole country. However, it is likely that if a similar study were conducted elsewhere in Finland the results would probably be similar and so would the interpretations. This is mainly due to the limited cultural differences in Finland: the Finnish population is rather homogenous in its structure if compared to larger European countries. Consequently, the data provides 'survey level' detail in the analysis of qualitative issues. Finally, the presented response frequencies do not reveal much about the underlying structures determining the obtained result. The empirical analyses of the structural processes that are beneath the observed results are a question for another study.

Results: technology access and social statements

First, I will discuss some general results regarding device availability. This means that the respondent is able to use and has access to the technology. The availability of mobile phones in the data was 95.9%. According to the official national statistics the corresponding figure for the same year was 95.5%. In this respect, the used data should reflect mobile communication relatively well. There were no significant gender differences in the availability levels of (mobile) technology except one: the availability of mobile Internet (access with a laptop computer) was considerably higher for men (16.1%) than for women (7.0%) resulting in a 9.1% difference.

Table 14.2 Selected claims from the data that have highly significant chi-square value (the connection to the framework is in bold above each claim)

Tools for communication (box 2, Figure 14.1.)

Claim		I like to write text messages (SMS)					Total
		Fully agree	Somewhat agree	Hard to say	Somewhat disagree	Fully disagree	
Gender	Women	211 (37.4 %)	154 (27.3 %)	61 (10.8 %)	84 (14.9 %)	54 (9.6 %)	564 (100 %)
	Men	69 (15.5 %)	111 (24.9 %)	75 (16.9 %)	117 (26.3 %)	73 (16.4 %)	445 (100 %)
Total		280 (27.8 %)	265 (26.3 %)	136 (13.5 %)	201 (19.9 %)	127 (12.6 %)	1009 (100 %)

Chi-square: 75.71 ***

Claim		Public sector should actively aim to promote IT skills within population					Total
		Fully agree	Somewhat agree	Hard to say	Somewhat disagree	Fully disagree	
Gender	Women	73 (12.6 %)	268 (46.3 %)	180 (31.1 %)	48 (8.3 %)	10 (1.7 %)	579 (100 %)
	Men	108 (23.8 %)	195 (43.0 %)	109 (24.0 %)	34 (7.5 %)	8 (1.8 %)	454 (100 %)
Total		181 (17.5 %)	463 (44.8 %)	289 (28.0 %)	82 (28.0 %)	18 (1.7 %)	1033 (100 %)

Chi-square: 23.55 ***

Tools and need for communication (box 3, Figure 14.1.)

Claim		I prefer to use a mobile phone rather than a fixed phone to contact others					Total
		Fully agree	Somewhat agree	Hard to say	Somewhat disagree	Fully disagree	
Gender	Women	204 (36.2 %)	122 (21.7 %)	86 (15.3 %)	100 (17.8 %)	51 (9.1 %)	563 (100 %)
	Men	182 (42.3 %)	122 (28.4 %)	54 (12.6 %)	50 (11.6 %)	22 (5.1 %)	430 (100 %)
Total		386 (38.9 %)	244 (24.6 %)	140 (14.1 %)	150 (15.1 %)	73 (7.4 %)	993 (100 %)

Chi-square: 19.28 ***

Claim		Work-related features of mobile phones are important to me (calendar, data transaction)					Total
		Fully agree	Somewhat agree	Hard to say	Somewhat disagree	Fully disagree	
Gender	Women	16 (2.9 %)	52 (9.3 %)	51 (9.2 %)	123 (22.1 %)	315 (56.6 %)	557 (100 %)
	Men	26 (5.9 %)	75 (17.1 %)	50 (11.4 %)	105 (23.9 %)	183 (41.7 %)	439 (100 %)
Total		42 (4.2 %)	127 (12.8 %)	101 (10.1 %)	228 (22.9 %)	498 (50.0 %)	996 (100 %)

Chi-square: 29.39 ***

Experience (box 4, Figure 14.1.)

Claim		I think I use the internet too much					Total
		Fully agree	Somewhat agree	Hard to say	Somewhat disagree	Fully disagree	
Gender	Women	3 (0.6 %)	5 (1.0 %)	25 (5.1 %)	102 (20.9 %)	354 (72.4 %)	489 (100 %)
	Men	5 (1.3 %)	17 (4.4 %)	54 (14.0 %)	129 (33.5 %)	180 (46.8 %)	385 (100 %)
Total		8 (0.9 %)	22 (2.5 %)	79 (9.0 %)	231 (26.4 %)	534 (61.1 %)	874 (100 %)

Chi-square: 66.10 ***

Claim		The mobile phone has increased my sense of security					Total
		Fully agree	Somewhat agree	Hard to say	Somewhat disagree	Fully disagree	
Gender	Women	234 (41.1 %)	259 (45.6 %)	49 (8.6 %)	19 (3.3 %)	7 (1.2 %)	568 (100 %)
	Men	129 (29.0 %)	197 (44.3 %)	78 (17.3 %)	23 (5.2 %)	18 (4.0 %)	445 (100 %)
Total		363 (35.8 %)	456 (45.0 %)	127 (12.5 %)	42 (4.1 %)	25 (2.5 %)	1013 (100 %)

Chi-square: 36.24 ***

Claim		24/7 mobility and instant personal 'availability' is stressful					Total
		Fully agree	Somewhat agree	Hard to say	Somewhat disagree	Fully disagree	
Gender	Women	28 (4.9 %)	100 (17.8 %)	92 (16.3 %)	195 (34.7 %)	147 (26.2 %)	562 (100 %)
	Men	36 (8.2 %)	96 (21.7 %)	106 (24.0 %)	121 (27.4 %)	82 (18.6 %)	441 (100 %)
Total		64 (6.3 %)	196 (19.5 %)	198 (19.7 %)	316 (31.5 %)	229 (22.8 %)	1003 (100 %)

Chi-square: 15.10 ***

However, there were no significant gender differences in computer and Internet usage times and frequencies. Some 55% of respondents used computers daily and the corresponding figure for Internet use was approximately 45%. The technology ownership and availability followed a combination of key background variables. These included education (elementary and professional), professional position and age.

Therefore, it is worth deepening the analysis to the level of classification of other key explanative variables of technology use. In this data, the respondents with lower education were similar to each other and gender differences were not significant, but the difference became significant in groups with higher education (academic and higher secondary education). Thus, educated women differed from their male counterparts on their levels of mobile Internet availability. If we consider age, the largest difference occurred in the age group of 50–60-year-olds. The age grouping reflects the fact that in time the significance of age in the explanation of technology use will diminish. The largest age gap was in the particular group of 50–60-year-olds. Younger cohorts (from 20-year-olds upwards) were rather similar in their technology usage levels and practices. Younger people (children and young people up to 20 years of age) on the other hand, differed greatly from the 'adult' age groups.

In general women tended to have slightly fewer devices than men. However, the differences were in most cases statistically not significant. A similar result has been obtained in the latest national survey of Statistics Finland (2006c). The differences in that study were less than 5% in general. The gender division of ICT mobility is thus not derived from the availability of and access to technology in Finland. The analysis of the availability of technology is a pragmatic and simple way to look at the differences of mobility. It is more interesting to look at distributions of experience and practice-related statements. In Table 14.2 the key results have been provided for the selected statements. The single statements have been organised in a manner similar to the framework (Figure 14.1) and contain the largest gender differences in the data.

Next I discuss some of the key elements found in the distributions. First, the greatest gender difference in the data was in the experience of writing SMS messages. This result supports earlier studies (Kasesniemi, 2003, 78–84; Thulin, 2004, 82–83). The data also indicated that women tend to have a more sceptical attitude towards the development of an information society in general. In addition, men experienced the information society as more work-oriented than women. This was reflected in the usage patterns of mobile technology. Data transaction tools for mobile devices were generally more 'important' to men than women. This issue is connected to work positions and gendered dynamics of work life in general. Interestingly, men also experienced more stress related to the constant availability associated with the 24/7 statement.

One of the most important issues in the gender differences for mobile telephone usage was related to safety. The issue of safety was closely connected to the experience of family communication between parents and their children. In a majority of cases, the technological development in mobile communication has extended possibilities for interaction. However, it should be acknowledged that the issue of communication content is not technology dependent. This general statement is supported by the data. The social practices and the related historical development that has shaped them are of the utmost importance. These social histories include, for example, household power relations that are also important; women tend to have more responsibility in household and childcare matters than men, which is reflected in communication patterns. These include taking care of children and making certain that children are safe when away from home. Mobile phones are an efficient device to foster these contacts.

There were no skill-related differences between women and men in the data. Both genders experienced their skills to be similar with respect to using mobile technologies. Only a marginal proportion of respondents felt unable to use mobile phones because of a lack of skill (1.7% of women and 1.5% of men). An interesting statistically 'non-significant' finding was that there was no gender difference in the experience of technologies improving ones' life. The statement used was: 'The Internet and email have improved my social life.' The full distribution of the statement was also surprising: only 19.7% of all respondents agreed fully or partially with the statement whereas those answering 'fully disagree' totalled 37.2%. Thus the common belief in technological bliss is a fragile thing. Technology does not automatically lead to better social conditions if the existing social networks and relations are not good. In other words, meaningful things in life are beyond technologies. Technologies are tools that can only enhance the practice of communication. They do not provide the contents for social interactions. This result can also be interpreted as a critique of technological determinism (for example Graham, 1998, 167–71) and the 'non-significance of the human actor' in the development of the information society.

To conclude this short empirical excursion I will make the following key interpretations. The gender differences in everyday usage patterns of technology are relatively small. However, the most significant differences reflect broader societal issues that transcend individual habits. These issues are connected to the experiences of personal and family safety, experiences of technology as a tool for work, experiences of working culture, and values associated with society.

An individual's work orientation shows a clear gender difference: men value and appreciate mobile technology more than women do. They are also more work-oriented in their mobile communications. One of the most important issues in Finland and other Nordic countries is the question of equality in working life. Statistics show that women have lower salaries in general and their presence in top management positions is rare. The work-oriented drive of highly educated men seems to be illustrated by this data. The gender differences tend to increase in the highest education levels. The result is important because some 60% of current university graduates in Finland are women. This is one issue for policy-makers to consider.

The second important issue is the experience of safety and of feeling safe. This result also has a long historical tradition behind it. The experience is an individual feeling (or interpretation) of a larger social phenomenon, connecting societal development, social coherence and equality and well-being that, combined, result in a safer society with low criminal rates. These individual practices and experiences can (and should) be connected to the macro-level policy guidance.

Conclusions: From policies to everyday practices

This chapter has drawn on an analysis of survey data and policy documents. The existence of the current gender system is based on a long history of societal development. The everyday practices of technology use have dimensions reflecting this system. Women use and appreciate different aspects of mobility. The observed differences in the sense of security, household communication between children and parents and differentiation in the method of communication are not just symbols of 'random choices' but reflect issues deeper in society. The role of technology itself becomes meaningless if the underlying values and history of societal development are not taken into consideration.

The chapter's main hypothesis concerning the invisibility of gender in information society strategies proved to be accurate. The issues that are presented daily in the media, such as gender equality, are easy to talk about but the actual implementation is much more difficult. The strategy papers are important documents to steer decision-making. In the case of the information society, the progress of gender issues has been slow.

The problem lies in the fact that the current methods of devising societal strategies are rather monolithic. This means that the preparation of policies and strategies relies strongly on a certain narrow type of expertise that stresses competitive success, an understanding of the indicators of economic success, the ability to rationalise one's views in the language of strategic aims, as well as the achievement of a competitive advantage and efficiency gains.

In fact, gender issues were even less prominent than I first assumed, with respect to the aspects that have been incorporated into the current Finnish information society strategy. The emphasis was largely focused on e-services and infrastructure: on the availability of mobile connections for individuals and services to use with the devices. This is not surprising because of the process of strategy creation: the strategies are based more on the voices of men than women. On the other hand, in the strategy text itself, gender issues are not taken into consideration. The strategy thus sees individuals as objects – targets of actions – not as active participants. Critiques from the social sciences have pointed this out for a long time but it is still evident that strategies, created by the elite, consider individuals as populations.

The dimensions of work-life conditions, social equality resulting in safer environments and appreciation of everyday practices, as well as recognition of 'small' things in the creation of policy guidance should be considered more thoroughly in the strategy process. A common statement put forward in connection with the information society and information society strategies is that survival within global competition is possible only through a combination of technological advancement, education and an efficient economy. The information society strategy process can be compared to a puzzle and the act of putting one together. The information society involves contemplating and understanding the interdependencies between the pieces of the puzzle and the factors that change, alter or affect the pieces. The above statement affords strategies an important position in the development of the information society and sets new demands for the future.

References

Abbot, P. and Sapsford, R. (1988), *Women and Social Class* (London: Tavistock).

Archibald, J. et al. (eds) (2005), *The Gender Politics of ICT*. Papers from the 6th international conference of women in computing (London: Middlesex University Press).

Barbee, A.P., Cunningham, M.R., Winstead, B.A., Derlega, V.J. Gulley, M.R. Yankeelov, P.A. and Druen, P.B. (1993), 'Effects of gender-role expectations on the social support process'. *Journal of Social Issues* 49:3, 175–190.

Chatrie, I. and Wright, P. (2000), *Public Strategies for the Information Society in the Member States of the European Union*. ESIS report. Available at <http://europa.eu.int/ISPO/esis/default.htm>, accessed 12 September 2006.

Cohoon McGrath, J. and Aspray, W. (eds) (2006), *Women and Information Technology: Research on Underrepresented* (Cambridge, MA: MIT Press).

Cox, E. and Leonard, H. (1993), *Weaving Community Links. The Cost Benefit of Telephones in Maintaining the Social Fabric Through the Unpaid Works of Women* (Sydney: Distaff Associates).

DeSolaPool, I. (ed.) (1981), *The Social Impact of Telephone* (Cambridge, MA: MIT Press).

Dimmick, J. et al. (2000), 'The gratification niches of personal e-mail and the telephone. Competition, displacement and complementaritity', *Communication Resarch* 27 (2), 227–48.

Evans, J. and Brooks, L. (2005), 'Understanding collaboration using new technologies. A structural perspective', *Information Society* 20 (3), 215–20.

FSD (2007), Information society in Pirkanmaa region 2004. <http://www.fsd.uta.fi/english/data/catalogue/FSD2010/meF2010e.html> Accessed 18.12.2007.

Government Policy Programmes (2006), Information Society Programme. <http://www.tietoyhteiskuntaohjelma.fi/esittely/en_GB/introduction/_files/11233297000000607/default/tietoyhteiskuntaohjelma_2006_en.pdf>Accessed 18.12.2007.

Graham, S. (1998), 'The end of geography or the explosion of the place? Conceptualizing space, place and information technology', *Progress in Human Geography* 22:2, 165–185.

Grint, K. and Gill, R. (eds) (1995), *The Gender Technology Relation. Contemporary Theory and Research* (London: Taylor & Francis).

Hannemyr, G. (2003), 'The internet as hyperbole: a critical examination of adoption rates', *Information Society* 19 (2), 111–22.

Haraway, D. (1991), *Simians, Cyborgs and Women. The Reinvention of Nature* (London: Routledge).

Haynes, D.J. et al. (eds) (2005), *Gender, race and information technology*, a special issue of the journal *Frontiers: A Journal of Women's Studies* 26 (1).

Hey, V. (1997), *The Company She Keeps. An Ethnography of Girls Friendship* (Buckingham: Open University Press).

Inkinen, T. (2003), 'Information society, citizens and everyday life: does the Internet make a difference in spatial practices?' *Fennia* 181 (1), 25–33.

Jauhiainen, J.S. (2007), 'Men, money and mobile phones? Tracing the technology discourse in the Finnish newspapers', *Technology in Society* 29:1, 79–91.

Kasesniemi, E.-L. (2003), *Mobile Messages. Young People and New Communication Culture* (Tampere: Tampere University Press).

Kellerman, A. (2006), *Personal Mobilitie* (London: Routledge).

Larsen, J., Urry, J. and Axhausen, K. (2006), *Mobilities, Networks, Geographies* (Aldershot: Ashgate).

Lieber, C.A. and Sandefur, G.D. (2002), 'Gender differences in the exchange of social support with friends, neighbors, and co-workers at midlife'. *Social Science Research* 31:3, 364–391.

May, C. (2002), *Information Society. A Sceptical View* (London: Polity).

Morgan, K. et al. (eds) (2004), *Human perspectives in the internet society: culture, psychology, and gender* (Southampton: MIT Press).

Mörtberg, C. (2003), 'Heterogeneous images of (mobile) technologies and services: a feminist contribution', *NORA: Nordic Journal of Women Studies* 11 (3), 158–69.

OECD (2005), *Territorial Reviews. Finland* (Paris: OECD).

SITRA (1998), 'Quality of life, knowledge and competitiveness: Premises and objectives for strategic development of the Finnish information society', *SITRA publications 211,* Helsinki.

Spender, D. (1995), *Nattering on the Net: Women, Power and Cyberspace* (Melbourne: Spinifex).

Statistics Finland (2006a), *Koulutus 2006:2 (education 2006:2)* (Helsinki: Statistics Finland).

Statistics Finland (2006b), *Väestö 2006:1 (population 2006:1)* (Helsinki: Statistics Finland).

Statistics Finland (2006c), *Suomalaiset tieto- ja viestintätekniikan käyttäjinä (Finnish people as users of information and communication technologies)* (Helsinki: Statistics Finland).

Thulin, E. (2004), *Ungdomars virtuella rörlighet. Användningen av dator, Internet och mobiltelefon i ett geografiskt perspektiv*. Publications edited by the Departments of Geography, Series B 105. (Göteborg: University of Göteborg).

Transparency International (2006), *Global Corruption Report 2006* (London: Pluto Press).

Turkle, S. (1995), *Life on the screen. Identity in the age of the Internet* (New York: Simon & Schuster).

Wajcman, J. (1991), *Feminism Confronts Technology* (Cambridge: Polity Press).

Wellman, B. and Wortley, S. (1990), 'Different strokes for different folks: community ties and social support', *American Journal of Sociology* 96:3, 558–588.

Chapter 15

Gender Mainstreaming in Swedish Transport Policy

Merritt Polk

Introduction

Sweden is currently putting a great deal of effort into discussions regarding how gender equality can be achieved in the transport sector. This is due to a political strategy that began in 2001, with the addition of a gender equality subsidiary goal to Swedish transport policy. In the five years since the goal was put into effect, there has been a great deal of discussion and debate regarding what a gender equal transport system entails, the relevance of the goal for the sector, and how it can be achieved in practice. However, despite such groundbreaking political support and the five years within which actions could have been initiated, there has been little concrete change in the distribution of power between women and men, with only limited support for research that aims at understanding the role of gender in the sector, and few concrete strategies to address gender equality. The government's recent proposition, *Modern Transport*, released in March 2006, has reaffirmed the government's commitment to gender equality with a new emphasis on the relation of gender equality to the other transport political goals and by renewed commitment to equal representation (Prop. 2005/2006:155).

In the Swedish context, the official goal of gender equality is 'a society where women and men have the same opportunities, rights, and responsibilities within all vital areas of life' (Skrivelse 1999/2000:24, 6). More specifically, gender equality refers to social structures and institutions that enable both women and men to successfully attain economic independence, individual fulfilment, and security, where both women and men have freedom from gender-based power structures, sexual violence, and discrimination based on sex. True and Mintrom (2001) define *gender mainstreaming* as 'efforts to scrutinize and reinvent processes of policy formation and implementation across all issue areas to address and rectify persistent and emerging disparities between men and women' (True and Mintrom, 2001, 28). Following from this, mainstreaming gender equality into transport policy should consider if and how transportation affects women and men and their ability to attain the overall political goals of gender equality. How, for example, do gender and transportation interact with an individual's ability to attain economic independence, to achieve personal fulfilment, and to feel safe and secure? Do the structures and institutions that make up the transportation sector support any gender-based power hierarchies or discrimination based on sex?

The overall aim of this chapter is to describe and analyse the current position of gender mainstreaming in the transport sector in Sweden. I will start with a history of legislation in Sweden on gender mainstreaming in transport policy. This will be followed by an analysis of why gender equality has not had more success, through the identification of a number of barriers to gender mainstreaming that currently exist in the sector. By way of conclusion, the consequences of the new legislation will be analysed in light of these barriers and applied to a discussion regarding the research that is needed to overcome obstacles to change.

A brief history of gender mainstreaming in the Swedish transport sector

The incorporation of a gender perspective in the transportation sector in Sweden began to take shape in 1994, when the Parliament adopted a national directive requiring that all committees and working groups include an analysis of how their proposals affect gender equality (Direktiv. 1994:124). The aim of this directive was to require the active mainstreaming of gender equality in all sectors of Swedish society. Within the transportation sector, this started within the overall process of designing a policy for a sustainable transportation system. In preparation for this, one of the expert groups convened in 1996 dealt specifically with gender equality. The 1997 final background report on transportation policy, *A new course in transportation policy*, included a chapter on gender equality (SOU, 1997:35). This chapter set the stage for gender mainstreaming within the sector by presenting differences in travel patterns between women and men and noting the lack of women in the decision-making process. Ensuing from this, the 1998 governmental proposition, *Transport policy for sustainable development*, resulted in the 1999 appointment of an advisory council charged with collecting information and suggesting measures for attaining gender equality within the transportation sector (Proposition 1997/98:56). This council was called the Gender Equality Council for Transport and IT. The Gender Equality Council presented their final report, *Gender equality – transportation and IT*, in June 2001 (SOU, 2001:144). The Gender Equality Council's 2001 report is a broad analysis that presents a history of gender equality, a vision for the future, methods for gender mainstreaming, and a summary of the laws and regulations that influence, or could be used to influence, gender equality (ibid). A clear connection is made between the goals of gender equality and sustainable development, especially regarding equal representation and participation in societal decision-making. A great deal of attention is also given to governmental measures that can be used to ensure gender mainstreaming, such as the use of socioeconomic modelling, public bidding, and requiring that companies that receive contracts from the state actively promote gender equality (ibid).

The culmination of the process of gender mainstreaming in transport policy occurred in October 2001, when the Swedish government approved gender equality as the 6th goal of Swedish transport policy in the governmental bill: *Infrastructure for a Long-term Sustainable Transportation System* (Proposition 2001/02:20). The bill stated: 'The transportation system shall be designed so that both women's and men's travel needs are satisfied; women and men shall be given the same possibilities to influence the system's design, formation and administration; and women's and men's values shall receive equal consideration' (ibid). The other subsidiary goals are a transport system that is (1) accessible, (2) effective, (3) safe, and promotes (4) environmental goals and (5) regional development.

From 2001 to 2006, there has been limited activity in the political arena. Overall, Swedish governmental agencies are steered through regulating directives that outline the agencies' goals, budgets and desired results. During this time, minimal directives for gender equality have been required such as a mapping out of differences in women and men's use of transportation and an account of the number of women and men who participate in public forums.

In March 2006, the government presented their latest legislation for transport policy (Prop. 2005/2006:155). In this bill, gender equality has become a required part of all other transport policy subsidiary goals, and there is a re-emphasis on the goal of equal representation in steering boards in the sector. Overall, this emphasis can be interpreted as a step towards a more thorough integration of gender equality. However, given the past problems in affecting change towards gender equality, the question must be raised if the present governmental strategy is adequate for overcoming the obstacles to gender equality that currently exist.

Current barriers to gender mainstreaming in the transport sector in Sweden

The last five years of having gender equality as a subsidiary goal are just the culmination of a longer political process. There is more than 25 years of gender equality legislation in Sweden and more than 10 years of gender mainstreaming by the government including, for example, specific gender equality directives for governmental committees. Even given years of political attention, the transport sector is far from gender equal. Many who work within the sector do not consider gender equality as being especially relevant or applicable. Many of the measures directed at gender equality in the sector are oversimplified and unorganised, and the resources allocated for gender research on transport issues are close to negligible. Given this situation, it is clear that there is a gap between sector politics and practice. The reasons behind why gender equality has been dealt with in such a lackadaisical manner must be recognised and fully understood to alleviate this problem. More legislation, such as the 2006 government bill, will not be successful unless the barriers to gender equality are clearly identified and thereafter applied to forming more efficacious political strategies.

In order to do this, it is first important to understand why the governmental agencies that have had the responsibility of implementing this legislation for the last five years have not made gender equality a priority. There are clearly many individual, political and structural factors that influence how the goal has been dealt with. The following will present a number of barriers that have been identified from an analysis of policy documents, formal and informal interviews with civil servants and politicians, research and consulting reports and participatory observation. The barriers that were identified are summarised in the following and dealt with in detail in the next sections:

1. Gender equality is not perceived as a relevant goal for the transport sector.
2. There is a lack of basic knowledge and expertise regarding gender and gender equality.
3. There is no systematic strategy for dealing with gender equality.
4. There is a lack of temporal and financial resources needed to deal with the goal.

1 Gender equality is not perceived as a relevant goal for the transport sector

Part of the criticism that has been directed at gender equality is that it is not a relevant goal for the transport sector. This can be explained by noting that the priorities, guidelines and goals within the sector are far removed from issues such as gender equality. Transport politics have prioritised more traditional goals such as adequate road capacity for both cars and trucks, minimising congestion, and more recently safety (as is exemplified in the overall goal of zero deaths from road traffic-related accidents), and the environment (in terms of reducing environmental degradation). Overall transport politics priorities have emphasised creating and maintaining a well-functioning transport system of high quality that is socio-economically effective. A common point that is brought up in discussions of gender equality is that the accessibility goal should subsume access for all members of the population, including women and men and that a specific goal for gender equality is redundant.

The transport sector in Sweden is also dominated by males in a number of gender-segregated occupations, such as those represented by technical and engineering fields. The sector is characterised by a certain approach to education, about working with and framing transport-related problems where gender equality represents an entirely different way of analysing and evaluating social interactions. In an analysis of the latest road traffic conference that was held in Göteborg in June 2006 – the Transport Research Arena (TRA) – it was found that road transport-related research is dominated by technical and instrumental perspectives and solutions, where the problem of attaining a sustainable transport system is predominantly framed in terms of technical solutions at the expense of more behavioural and political ones (Polk, 2006).

Furthermore, there is much research that shows that masculinity and technology are much more tightly connected than femininity and technology (Cockburn and Dilic, 1994; Creager et al., 2001; MacKenzie and Wajcman, 1999; Mellström, 1995; Wajcman, 1991). It can even be suggested that there are different subcultures within the sector (the automotive industry, car repairs, professional drivers, traffic administrators), where women and men are perceived as having different status and are associated with different types of meanings in relation to movement, distance, speed, engine power, cars, motors and trucks. Because women's sphere of expertise is not seen as being appropriate in male-dominated areas, this can, and often does, result in women being excluded scientifically, professionally and practically. An important barrier to gender mainstreaming is thus the domination of not only men, but also of worldviews and traditional priorities and educational paradigms where certain subjects, such as power relations between women and men, are inherently seen as irrelevant to the workings of the sector. They are therefore excluded from what is problematised or worthy of problematisation within the sector itself.

2 There is a lack of basic knowledge and expertise regarding gender and gender equality

The second and perhaps most important barrier to gender mainstreaming in the transport sector can also be seen as an explanation for the first, namely that there is a lack of basic knowledge and expertise regarding gender analysis in the sector. *Gender* refers to socially constructed differences between women and men. It is a property of individuals, but not the only one. As Robert Connell notes, '…we can speak of gender as collective and social practices as gender-structured' (Connell 1987, 139). Specific examples are found in traditional stereotypes surrounding definitions of masculinity and femininity as well as in the roles women and men have in the home, in how the media portrays women and men, in who does what jobs, and in who holds power in the private and public sectors. Furthermore, gender does not work alone; it is embedded in and works through many other social categories such as age, education, class, and ethnicity. It is an all-encompassing concept that permeates many different social factors and processes such as individual behaviours, structural organisations, and the meanings and symbols manifest in everyday life.

Gender equality is a complex societal goal that is based on a wealth of research predominantly from within gender studies and feminism. A vast amount of knowledge and expertise is needed both to understand and identify the problems underlying the lack of gender equality and to suggest measures for attaining more gender equality. Professionals within the sector are experts within their various fields of activity. However, educated women and men with years of practical experience do not necessarily have the knowledge necessary to understand and implement gender mainstreaming. One of the main reasons for this, as mentioned above, is that gender equality is not a part of the traditional fields and educational programmes that dominate the sector. This situation is changing, as governmental agencies in the sector are now requiring education in gender equality. However, despite such steps, it must be remembered that gender equality is also perceived as a controversial societal goal, especially in Sweden, which is seen as one of, if not the most, gender-equal countries in the world. Many question the need for continued emphasis on a goal that, for many, has already been achieved and where the importance of its position is seen as being overrated.

Another problem is the lack of applied and basic research regarding differences between women and men in the sector. Much of the research that is produced within gender studies is of a theoretical nature or specifically related to other empirical areas that are hard to extrapolate to the transport sector. Within the sector itself, research is carried out in mapping out concrete differences between women and men, with little incorporation of gender as a theoretical and analytical tool (Polk, 2005;TRANSEK, 2005), though there are exceptions to this (Andersson, 2005; Friberg et al., 2004; Polk, 2005). There is a decided need for research in the sector that combines both a theoretical and empirical perspective that can be used as a basis upon which more effective strategies and measures can be developed.

Gender mainstreaming is a strategy that is based on the assumption that there are structural barriers in society that maintain unequal power relations between women and men. The basic premise is that in order to break down such barriers, gender equality must be integrated in all areas of decision-making and practical application, which is a basic platform of the government's gender equality legislation. If such premises are not accepted, for example that there are structural barriers to gender equality, then gender mainstreaming itself can be questioned as a viable or even relevant strategy.

A number of specific debates have also arisen regarding the intersections of gender equality and transportation. One of the reasons that gender equality has not been seen as a relevant goal is due to confusion surrounding the role of the transport sector in attaining gender equality in the rest of society. For example, an important question that has been raised is if gender equality can be improved by changes made within the transport sector (Vägverket, 2005). There are a number of different positions with regard to this question. In the first, gender inequalities are seen as a societal problem that originate from sectors other than the transport sector and therefore should be solved in more applicable areas such as within the Ministries of Education and Research, Employment and Gender Equality, Democracy and Human Rights, and Health Care and Social Issues.

In the second position, the transport sector is seen as a tool for gender equality by improving women's and men's access to societal resources in the form of employment and leisure activities.

> The intersections between gender equality legislation and transport politics means that the transport sector should be an instrument for women and men to live equally – that it answers their transport needs... A gender equal transport system should facilitate women's as well as men's daily lives, increase their possibilities in obtaining work, studying or participating in inspirational leisure activities.
>
> Ulrica Messing, Minister of Infrastructure (Vägverket, 2005, 110, 2)

In this position, gender mainstreaming in the sector is seen as a vital link to increasing women and men's ability to attain gender equality and therefore a vital part of gender equality politics.

The third position shows the complexity and difficulty of integrating gender equality in the sector. Gender equality is seen as important – but only in limited areas. In goal formulation, gender equality is specifically directed towards three areas, namely, travel needs, values and representation. The part of the gender equality goal that deals with equal representation between women and men in decision-making and planning is rarely or never questioned as a viable goal for the sector. However, a great deal of discussion has occurred with regard to differences in travel needs between women and men and how gender equal travel patterns would look.[1]

---

1 In summary, both women and men travel 69% of their kilometers by car. The main difference between women and men is in the number of kilometers covered. Per day this is, on average, 55 for men and 35 for women. Regarding trip purpose, men travel to a much greater extent for work-related trips especially in the category of business trips, where they represent a majority of kilometers and trips, while women make a larger number of service-related

Striving for equal travel patterns between women and men is questioned because women's current travel patterns are seen as being based upon rational and voluntary prioritisations of the home, children, and proximity to the workplace, and therefore a logical explanation for the differences that currently exist. For those who see the differences in women and men's travel as an expression of inequality, improving women's present travel patterns, such as prioritising public transportation, is seen as a way of cementing current inequalities. There is thus no consensus on what gender-equal transport entails with regard to travel needs, or what specific indicators could be used to measure the attainment of such a goal.

It is clear that the sector must strive to eliminate and not cement differences between women and men's travel patterns, and that it should also avoid a stereotypical approach to gender equality, but these are not adequate reasons to turn the question of a gender-equal transport system into a problem that has no answer. It is important instead to gain a better understanding of the differences between women and men's travel patterns and to identify the driving forces behind these differences, which are a combination of individual choices, infrastructure, values, experiences and gendered power relations both in the home, the workplace and in society in general.

3 There is no systematic strategy for dealing with gender equality

Swedish politicians have possibly done more than any other country in the world regarding the promotion of a gender-equal transport system. But there are still shortcomings regarding how gender equality has been integrated into the procedures that comprise the Swedish political process. There are a number of clear measures regarding political procedures for implementing policy goals that were not taken advantage of, many of which were clearly outlined in the report from the Gender Equality Council in 2001. In a governmental communication on gender equality from 2002, the importance of a systematic approach to gender equality was made clear:

> Gender mainstreaming involves an increased responsibility for gender equality politics for all governmental agencies. Those who have responsibility for an agency are also responsible for integrating a gender perspective in their activities. This strategy is necessary and has been shown to be successful but it can only work if the responsibility for gender equality is combined with clear goals, knowledge and support. This isn't always the case today, which has resulted in that the gender equality work that is carried out in governmental agencies is often isolated, ad-hoc and not integrated in the regular functioning of the agency. The quality of working with gender equality must be increased. (Reg. skriv. 2002/03:140)

---

and shopping trips. Men make a majority of their trips as the driver of a car, while women divide their trips and kilometers as a passenger in a car, public transportation and walking. Women and men spend the same amount of time travelling per day, but men by faster modes. Differences between women and men's daily car use remain after controlling for car access in the household, employment levels, income, education and age (Krantz, 2000; Polk 1998b). (For summaries in English see Polk, 1998a, 1998b and 2004.)

The clearest measure for ensuring the integration of gender equality is the yearly governmental directives that steer the work that is done in all Swedish governmental agencies. Despite the suggestions made in 2001, the government has not integrated gender equality more than superficially as a specific requirement in the yearly reports in the transport sector. The lack of concrete guidelines and requirements has only added to the confusion. In the five years since the goal was adopted, the only requirement for sex-segregated statistics was with regard to participation in public forums. In 2005, a requirement for showing differences in the number of women and men who are killed in traffic-related accidents was added. There is no requirement for sex-segregated statistics regarding serious injury, seat belt use, driving under the influence, and speeding, even though all of these categories show large differences between the sexes.

A consequence of this unsystematic implementation of gender equality is that the responsibility for gender equality has been taken by a number of engaged individuals, often without extra resources. The work that has been achieved within the sector has been based on a few engaged and active individuals, and is not the result of an overall strategy. Thus gender equality, while clearly a topic of discussion, has been dealt with and seen as a special interest issue.

4 There is a lack of temporal and financial resources needed to deal with the goal.

This lack of a systematic strategy has led to the last barrier, namely a lack of financial and temporal resources for working with gender equality. The responsibility for gender equality has often been given to engaged individuals, as described above, or to those who are responsible for working with other subsidiary goals, most often without extra resources. One of the most important barriers to gender mainstreaming is the lack of employment positions directly connected to the goal, and a lack of money for basic and applied research. Even if the most recent legislation says that such research must be prioritised, in reality this is purely rhetoric. There are also demands that research projects in the sector must integrate gender equality, but even there the lack of competence is striking. Oftentimes research money is given to consulting firms that have no gender expertise whatsoever. Such an application follows clearly from the other barriers outlined above. Gender equality is not seen as an area that warrants in-depth analysis or one that demands or warrants competence and expertise.

A result of this view is a clear lack of knowledge and insight regarding how gender can be integrated in research in the transport sector. Researchers and consultants pay lip service to gender equality but the topic is not taken seriously. It is not seen as a priority, either for research or for funding. A number of strategies have been developed to counteract this trend, such as a working group that deals specifically with how gender equality can best be integrated in research funded by the different governmental and funding agencies. There are also, as mentioned earlier, a number of educational programmes that are being required for all personnel in leadership positions in some of the governmental agencies in the sector.

Effective strategies for the future

Given this five-year-long history of ineffectual gender mainstreaming in the transport sector, and the barriers outlined above, the question needs to be raised if the former government's latest strategy for gender equality will have more success in the future. On the one hand, it can be argued that five years is far too short a time to achieve a more gender-equal transport sector and that more time will show success. On the other, the question must also be raised as to how the barriers outlined above can be dealt with in practicality. What needs to be included in future strategies for attaining more gender equality in the transport sector in Sweden?

It is clear that there is a decided lack of knowledge regarding the role of gender in the sector. While a number of studies have been undertaken regarding differences in women and men's travel patterns, values and political representation, none of these includes a theoretical application of gender as an analytical tool (Polk, 2005; TRANSEK, 2005). Until the root causes of gender inequalities within the transport sector are understood, we will be unable to design adequate political measures for their alleviation. This is where the need for a combination of theory and practice comes to the forefront.

There are two main theoretically and empirically distinguishable issues raised by the gender equality goal. The first emphasises the gendered distribution of power and influence in the public sector. The success of this emphasis, it is reasoned, will ensure that concrete differences between women and men will be thoroughly integrated into all decisions and planning that is undertaken with the equal participation of women and men . However, even given equal participation, the classic liberal approach to gender equality has faced well-known limitations, such as glass ceilings and other types of limited access to power and influence, as is exemplified in the Swedish case. All in all, ensuring an equal number of women and men increases, but does not ensure, women's equal participation (Polk, 2004). It is therefore important to separate gender distributions of power from the inequalities between women and men that result from their unequal status and position in society. In other words, within transport policy, structural and organisational distributions of power must be analysed and dealt with separately from differences in individual needs and values.

The overall theoretical areas that can be used to better understand the role of gender in the sector lie at the intersection of gender analysis, gender equality, and studies of technology, masculinity, and sustainability. The goal is to understand the practical repercussions that the masculinity/technology black box has in the empirical context of the transport sector, and analyse what impact this has on achieving gender equality. There are three main problems or gaps in the theoretical and empirical research that need to be rectified in order to achieve this end.

First, although there are many feminist studies of science and specific feminist studies of different types of technologies, there is no clear theoretical framework that adequately integrates the study of gender with the study of technological fields of knowledge production and practical application, and the implications this has for the current governmental focus on sustainability (Cockburn, 1983; Cowan, 1983; Harding, 1986; Keller, 1985; Mellström, 1995; Wajcman, 1991). Science and Technology Studies is the research field where the most attention has been given to scrutinising the 'social shaping' or 'social construction' of socio-technological knowledge systems, networks and/or artefacts (Bijker and Law, 1991; Fogelberg, 2000; Latour, 1987); however, the conceptual frameworks developed therein disregard both distributions of power between women and men and their impact on the development of technological artefacts (Fogelberg, 2000; Law, 1991; Lohan, 2000). There are a number of frameworks regarding how to better integrate feminist studies with science and technology studies, but few recent examples of their empirical applicability, and none that are directed at the transport sector (Lohan, 2000).

There is thus a clear scarcity of theoretical work that can be applied to analysing how gender interacts with different dimensions of the transport sector. Masculinity researchers find a plurality of different constructions and combinations of what it means to be masculine, not only in different cultures and times but also within the same culture, suggesting that any attempt to dichotomise stereotypical gender relations will fail to capture this complexity (Connell, 1995; Mellström, 2002). Thus, even if technology, or technological spheres of activity, are often conflated with masculinity, or seen as being constitutive of masculinity, this relationship cannot be seen as given and must be analysed as dynamic and varied within empirical contexts with a focus on specific case studies.

Second and closely connected to the first, there is no empirical research on the gendered nature of the scientific and professional paradigms in the transport sector which would allow us to see if there is a gender bias regarding the paradigms or perspectives that are currently hegemonic and those that are excluded, and how the technical focus of the sector forms or interacts with other alternative approaches, such as integrating the goals of sustainable development. A majority of the research conducted in the 1980s and 1990s shows that men have a dominant role in shaping and producing technology (MacKenzie and Wajcman, 1999; Traweek, 1988; Wajcman, 1991). One study dealing specifically with the automobile industry suggests that 'the world of technology is a secure domain of social activity (which) expresses and consolidates relations among men' and 'if technology and technical skill are constitutive features of what it means to be masculine, the relation between femininity and technology is typically portrayed as one of non-existence' (Mellström, 1995, 169, 171). Applying such results to gender mainstreaming in the transport sector suggests a strongly gendered bias due to the technical orientation of the scientific and professional paradigms and perspectives resulting in the potential exclusion of women and alternative perspectives. The analysis of women's roles becomes an analysis of to what extent, how and why women are excluded or included. Gender, in the sense of the presence of women in an arena dominated by men, is thus transformed into an analysis of how a male-dominated area of activity is informed by gender and how this obstructs the participation of women.

The third theoretical area that must be included targets a more individual and practical level of analysis. It deals with both the question regarding if women in the transport sector represent another approach to framing problems that is explicitly different from their male counterparts, and if women in the sector would be interested in taking more responsibility for decision-making and planning. Regarding the former, many studies suggest that women have attitudes and experiences, especially regarding the environment, that differ significantly from men (Bennulf, 1994; Davidson and Freudenburg, 1996; Hunter, Hatch and Johnson 2004; Johnson-Latham, 2006; Polk, 2003, 2004; Zelezny, Chua and Aldrich, 2000). In a Swedish survey, women have been shown to be more supportive than men of inclusive and environmentally friendly priorities in the transport sector (Polk, 2005). Even though these research results suggest significant gender differences, the specific nature and expression of these differences has only been tested in a limited number of studies, a majority of which are dated. Consequently, assuming that more women in the transport sector will automatically increase gender equality in actuality is purely conjecture at this point. Given such a potential gender bias in such a highly male-dominated sector, having more women represented does not guarantee any increase in power or influence, nor is it clear if professional women hold attitudes that are more supportive of sustainable solutions than their male counterparts. There is recent research from England that suggests that women's preferences are the main cause of both horizontal and vertical segregation in the job market (Hakim, 2006). Thus the question can also be raised if there are other, more individual, barriers to attaining gender equality that are not currently integrated in the latest governmental strategy.

Without more empirical research based on existing theoretical frameworks regarding the specific role of gender in the transport sector, there is no way to understand the past failure of the sector to deal with gender mainstreaming, to evaluate the potential efficacy or weaknesses of the present bill, nor any basis upon which to suggest more efficient alternatives. The problems surrounding gender mainstreaming have to be correctly identified and understood before they can be solved.

References

Andersson, B. (2005), *Risk: Om kvinnors erfarenhet och fysisk planering*, Centrum för kommunstrategiska studier (Linköping: Linköpings universitet).

Bennulf, M. (1994), *Miljöopinionen* (Lund: Universitetsförlag Dialogos AB).

Bijker, W. and Law, J. (eds) (1991), *Shaping technology, building society: Studies in sociotechnological change* (London: MIT Press).

Cockburn, C. (1983), *Brothers: Male dominance and technological change* (London: Pluto Press).

Cockburn, C. and Furst-Dilic, R. (eds) (1994), *Bringing technology home: Gender and technology in a changing Europe* (Buckingham: Open University Press)

Connell, R. (1987), *Gender and power* (Cambridge: Polity Press).

Connell, R. (1995), *Masculinities* (Cambridge: Polity Press).

Cowan, R. (1983), *More work for mother: The ironies of household technology, from the open hearth to the microwave* (New York: Basic Books).

Creager, A., Lunbeck, E., Schiebinger, L. (eds.) (2001), *Feminism in 20th century science, technology and medicine* (Chicago: University of Chicago Press).

Davidson, D. and Freudenburg, W. (1996), 'Gender and environmental risk concerns. A review and analysis of available research', *Environment and Behavior* 28 (3): 302–39.

Direktiv. (1994:124), Kommittédirektiv. *Direktiv till samtliga kommittéer och särskilda utredare att redovisa jämställdhetspolitiska konsekvenser*, Beslut vid regeringssammanträde 1994-11-10.

Fogelberg, H. (2000), *Electrifying visions: The technopolitics of electric cars in California and Sweden during the 1990's*. PhD dissertation (Göteborg: Science and Technology Studies. Göteborg University).

Friberg, T., Brusman, M. and Nilsson, M. (2004), *Persontransporternas vita fläckar, om arbetspendling med kollektivtrafik ur ett jämställdhetsperspektiv* (Linköping: Linköpings universitet).

Hakim, C. (2006), 'Women, careers, and work-life preferences', *British Journal of Guidance and Counselling* 34 (3), 279–94.

Harding, S. (1986), *The Science question in feminism* (Ithaca: Cornell University Press).

Hunter, L., Hatch, A. and Johnson, A. (2004), 'Cross-national gender variation in environmental behaviours', *Social Science Quarterly* 85 (3), 677–95.

Johnson-Latham, G. (2006), *Do women leave a smaller ecological footprint than men? Initial study of lifestyle, consumption patterns, sustainable development and gender* (Stockholm: Report from the Swedish Ministry of Sustainable Development, April 2006).

Keller, E. (1985), *Reflections on gender and science* (New Haven: Yale University Press).

Krantz, L.-G. (2000), *Rörlighetens mångfald och förändring: Befolkningens dagliga resande i Sverige 1978 och 1996*. Doctoral dissertation (Göteborg: Handelshögskolan).

Latour, B. (1987), *Science in action* (Cambridge: Harvard University Press).

Law, S. (ed.) (1991), *A sociology of monsters: Essays on power, technology and domination* (New York: Routledge).

Lohan, M. (2000), Constructive tensions in feminist technology studies, *Social Studies of Science* 30 (6), 895–916.

MacKenzie, D. and Wajcman, J. (1999), *The social shaping of technology* (Oxford: Open University Press).

Markör A.B. (2004), *Representation av kvinnor och män I beslutande organ inom Transportsektorn: En kartläggning på uppdrag av SIKA, Banverket, Luftfartsverket, Sjöfartsverket och Vägverket* (Örebro: Markör AB).

Mellström, U. (1995), *Engineering Lives. Technology time and space in a male centered world*. PhD dissertation (Linköping: Linköping University).

Mellström, U. (2002), 'Patriarchal machines and masculine embodiment', *Science, Technology & Human Values* 27 (4), 460–78.

Polk, M. (1998a), 'Swedish men and women's mobility patterns: Issues of social equity and ecological sustainability', in *Women's travel issues: Proceedings from the second national conference October 1996*. Washington DC: U.S. Department of Transportation, Federal Highway administration, Office of Highway Information Management. Pub. No. FHWA-PL-97-024.

Polk, M. (1998b), *Gendered Mobility: A study of women and men's relations to automobility in Sweden*. Humanekologiska Skrifter 17. Doctoral dissertation (Göteborg: Department of Interdiscipinary Studies, Section of Human Ecology).

Polk, M. (2003), 'Are women potentially more accommodating than men to a sustainable transportation system in Sweden?' *Transportation Research Part D*, 75–95.

Polk, M. (2004), 'The influence of gender on daily car use and willingness to reduce it in Sweden', *Journal of Transport Geography* 12 (2), 185–95.

Polk, M. (2004), 'Gender mainstreaming in transport policy in Sweden', *Kvinnor, Kön och Forskning*. 13 (1), 43–54.

Polk, M. (2005), 'Women's and men's valuations of road system infrastructure in Sweden', Vägverket internal report (Borlänge: Available at Vägverket's web pages).

Polk, M. (2006), 'The sustainability of the road transport research agenda in the European Union', Paper submitted to the TRB 20067 conference.

Proposition (1997/98:56), *Transportpolitik för en hållbar utveckling* (Stockholm: Riksdagens tryckeri).

Proposition (2001/02:20), *Infrastructure for a Long-term Sustainable Transportation System* (Stockholm: Riksdagens tryckeri).

Prop. (2005/2006: 155), *Moderna Transporter* (Stockholm: Riksdagens tryckeri).

Reg. skriv. (2002/03:140), *Jämt och ständigt – Regeringens jämställdhetspolitik med handlingsplan för mandatperioden* (Stockholm: Riksdagens tryckeri).

Skrivelse (1999/2000:24), *Jämställdhetspolitiken inför 2000-talet* (Stockholm: Riksdagens tryckeri).

SOU (1997:35), *Ny kurs i trafikpolitiken: slutbetänkande av Kommunikationskommittén* (Stockholm: Norstedts Tryckeri AB).

SOU (2001:44), *Jämställdhet – transporter och IT*. Slutbetänkande från Jämit – Jämställdhetsrådet för transporter och IT (Stockholm: Elanders Gotab AB).

TRANSEK (2005), *Jämställdhet vid val av transportmedel*, (Stockholm: TRANSEK).

Traweek, S. (1988), *Beamtimes and lifelines. The world of high energy physics* (London: Harvard University Press).

True, J. and Mintrom, M. (2001), 'Transnational networks and policy diffusions: The case of gender mainstreaming', *International Studies Quarterly* 45, 27–57.

Vägverket (2005:110), Res Jämt- tankar kring ett jämställt transportsystem (Borlänge: The Swedish Road Administration).

Wajcman, J. (1991), *Feminism confronts technology* (Cambridge: Polity Press).

Zelezny, L., Chau, P. and Aldrich, C. (2000), 'Elaborating on gender differences in environmentalism', *Journal of Social Issues* 56 (3), 443–57.

Chapter 16

Are We There Yet? Women and Transport Revisited

Clara Greed

Introduction

This book has discussed the journeys of people, goods, and ideas. However, most people equate mobility with physically travelling from A to B. Nowadays, the majority of people's journeys take place in crowded, busy, urban, and often fraught, physical surroundings. This is particularly true for women. Therefore, this chapter focuses upon the challenges of integrating a gender perspective into urban spatial transportation planning, in a world in which transport systems and their design, modes of transport available, and policy priorities continue to embody a male, rather than female, perspective.

The chapter is illustrated by examples predominantly drawn from the Anglo-American tradition of urban planning, including Britain. The discussion is applicable to any environment where 'the car is king': where the private car has become the main form of mass transport. Public transport alternatives have been increasingly marginalised and are, therefore, in practical terms, not viable practical alternatives for many sectors of the population, particularly those living in lower density, suburban residential areas, which comprise over 70% of available housing in Britain and around 80% in North America (Kunstler, 1993).

After many years of promoting the automobile, planners have made a *volte-face*, to condemn car use in the name of sustainability. This chapter contends this has not improved the situation for women, because the new generation of green transport planners has not taken into account the differential gender impact of their policies. It will challenge the shibboleth of 'sustainability', which should not be seen as above reproach and or beyond criticism. This chapter will be deemed politically incorrect and bourgeois by some readers, because of the emphasis on 'selfish' issues such as the school run, road charging, childcare, and the state of public transport infrastructure. But the chapter seeks to do justice to ordinary women and their concerns, their daily struggles to get to work, and their practical realities, all of which have been frequently voiced in the course of a range of recent research activities (Hamilton, 2000; Greed, 2005). In an ideal world, a 'sustainability' perspective would fully take gender considerations into account in all aspects of urban policy and transportation planning. The original definition of sustainability accepted with the Rio Declaration (1992) (Buckingham-Hatfield, 2000) included three components: economic viability and social equity, as well as environmental sustainability (that is prosperity, people, place). But, in Britain, a 'peopleless' concern with the 'green' environmental dimension of sustainability has often eclipsed the social component, and detracted from gender priorities and women's needs. If one says something to this effect, one may be told, 'oh we've done women, you should be concerned about the environment'. Yet many of the basic, and longstanding demands to address women's needs have not been addressed, let alone recognised within the new, sustainable approach to land-use transportation planning. Meanwhile in the social policy arena, gender has just become but one equality issue among many, within the new diversity paradigm (ODPM, 2005). But, gender is 'one of the most pervasive classificatory principles, arguably the most pervasive with a myriad of implications' (Kirton and Greene, 2003, 9). Gender cross-cuts and overarches all other social groupings (Greed and Reeves, 2005). Women comprise the majority of many minority groupings, as in the case of the elderly, people with disabilities and those dependent on public transport (ONS, 2005).

Admittedly, some battles have been won. It is now acceptable to write about 'women and transport' and 'gender and urban planning'. A veritable 'gender industry' has developed, with numerous academic outputs, conferences, consultancies, policies and recommendations. But, upon closer inspection, one finds the battle for policy implementation is still being fought. Rather than women and their needs being recognised within the mainstream of male academic activity and government policy-making, there are now two sets of conferences, reports, academic literature, one for women and one for men. Men, and their priorities, still dominate the bulk of transport policy-making, urban spatial policy and professional education, with few notable exceptions. Just look at the list of speakers at any major conference on transport or the environment or check the gender ratio of senior policy-makers in the built environment professions (De Graft-Johnson et al., 2006).

Therefore this chapter will seek to illustrate the fact that gender considerations still need to be recognised and addressed within transportation planning circles, so 'we are not there yet', by any means. The first section summarises 'the problem', created by traditional approaches to urban planning and transportation policy. The second section discusses the extent to which women's needs are being taken into account within a changed policy context in which both environmental and equality issues predominate. In the final section, as a solution to the present impasse, gender mainstreaming is introduced as means of integrating women's needs into transportation planning.

What is the problem? The city of man

Although 'planning is for people', it has been shown by research and human experience that women and men still have different roles and responsibilities. Thus it follows that they have different travel patterns within the city of man and so different policy needs (Greed, 1994; Anthony, 2001). Women are still predominantly the ones responsible for childcare, shopping and home-making, although the majority of women also work outside the home too. This means that women's 'journey to work' is likely to be more complex and multi-purpose compared with the simple 'home to work' and back again journey of the traditional male commuter. In contrast, a woman may set off from home, stop off at the childminder, then school, get to work, and return via the school gates, shops and childminder, resulting in complex trip chaining. This daily travel itinerary is difficult to achieve when employment and residential areas have been separated out by traditional land-use zoning, based upon male perceptions of functionality (Booth et al., 1996).

Arguably, urban policy has made the situation worse than it might have been without any town planning intervention, by prioritising housing clearance and the relocation during the post-war period of housing, industry and public facilities; by endorsing low density, suburban development around our cities; by investing in motorway and car-based infrastructure and letting public transport decline, and subsequently by permitting a whole range of types of development, including out of town shopping malls and hypermarkets, to develop alongside the new motorway system, thus undermining the viability of existing city centres and traditional towns in the process. All these policies have increased traffic congestion and journey times as people try to travel between different land-uses to carry out their daily tasks. It is now a virtual necessity to possess a motor car, particularly in cities where public transport has been severely cut back. Women's journeys are poorly met by public transport systems that have been designed on a radial basis funnelling workers into the centre during the rush hours, and provide limited off-peak services for women part-time workers, and a lack of transverse, inter-district bus routes. Trip-chaining, and multi-tasking are key features of women's travel, and an inevitable result of trying to combine their home and work duties. Women travel around cities using their survival skills and precise knowledge of local bus timetables, with very limited time budgets to get everything done.

Women's car and bus journeys, such as vital food-gathering shopping trips and child-escort journeys, have often been ignored, undervalued, or portrayed as 'fun' or as 'leisure' trips rather than 'essential' journeys and thus have been perceived as of little importance to public transport operators. This disregard for women's journeys was starkly revealed during a 'fuel blockade' undertaken nationally, by truck drivers protesting against rising prices in 2003. Emergency petrol supplies were made available to ambulances and essential staff. But hospitals soon found that many of their nurses, cleaning, catering and caring staff could not get into work as there was no petrol for their private car journeys. Clearly, such women were not seen as essential workers and barely figured in emergency planning statistics. There are half a million care workers nationally and 90% are female (Gilroy and Booth, 1999). More people are involved in catering and hospitality industries than there are in heavy industry. Traditional working class male factory jobs are in the minority nowadays, but still they are seen as more 'important' by policy-makers (ONS, 2005). The Greater London Authority (GLA) has put great emphasis upon equality monitoring (GLA, 2003a, b), but it still has blind spots. A recent decision to cut early Tube trains on the London Underground has seriously affected a whole army of cleaners, generally comprising women, recent immigrants and unskilled workers, who rely on this service but whose value and presence remains invisible to policy makers. For more local journeys, walking remains a major component of women's travel (GLC, 1985). But until relatively recently, walking did not count as transport. According to the Department of the Environment, 'transport is something you get into' (shoes or bicycles?). In the past, government surveys did not count journeys under 1 mile as valid for transportation planning purposes (1 kilometre = three fifths of a mile).

The Government response

In spite of all these gender-specific differences, women's voices are seldom heard, particularly within a transportation culture in which all motorists are seen as 'bad', and culturally, additional venom is reserved for 'women drivers'. Typically, the modern 'green' male environmentalist may justify his disdain for women as lazy and selfish with as much self-righteousness as the old-time 'red' trendy Marxists back in the 1960s used to condemn the early feminists for being 'bourgeois and trivial' in their demands for equality. As I struggle along through a pedestrianised part of my local city centre carrying tons of bags to a meeting far from the car park, I often encounter a young man, dressed up in the latest fashionable 'lycra cycling gear' coming up silently behind me and then whooshing past aggressively, cursing at me under his breath for daring to be on a footpath (presumably because he is purer than I because he is a cyclist) without so much as a warning tinkle on his bell. I wonder about him. Who does your shopping? Who does your washing? Who is looking after your children? (Presumably his mother, girlfriend, or wife.) I often encounter such young men at planning conferences who have taken up a career in environmental planning. Their policies are no doubt influenced by their life experience. They are likely to have no family responsibilities, mainly eat out, or often live in a rich little gentrified, inner-urban enclave near to where they work, where the local shops and amenities have not yet closed and public transport still exists and who have no idea how everyone else lives. Twice as many men as women cycle to work in Britain (Reeves and Greed, 2003). Deterrents include abuse from male drivers, and personal safety fears on cycle paths, as well as more conventional road safety concerns. So a gender lens needs to be applied to all forms of transport, and a greater emphasis needs to be given to qualitative (why and who) factors as well as quantitative data (how many) in determining transport policy. Unless they are constantly 'reminded' by women's groups, policy-makers will miss a whole layer of information resulting in a poor fit between planning policy and everyday life (DTI, 2005).

Nationally, the British government has employed a range of car-parking restriction measures, road charging, restrictive traffic management systems, and petrol price increases to discourage people from using their cars. But many people have no alternative; in the provincial towns, suburban residential areas and rural locations, public transport systems remain limited. Penalising the motorist does not give individual citizens the power to reopen railway stations or to move shops back into the centre. Ironically many women have actively protested against station closures and negative policy-decisions, and so can hardly be blamed individually for taking to the car – as a last resort.

National planning policy guidance has sought to restrict car parking to one space per house, and in the name of sustainability and better urban design (ODPM, 2003) car-free neighbourhoods are actively promoted by single male planners with little understanding of the complexity of daily family life (Early, 2006). This has been seen as discriminatory in preventing women using a car to get to work. Women are not free agents in choosing where to live or work, as they often 'follow' their husband's career and have to travel considerable distances to find work appropriate to their education and experience. The gender implications of parking standards need to be considered more carefully, particularly in new housing areas where no alternative bus route is available, and there is a lack of local facilities.

Increased parking charges in central areas, workplace parking charges, and reductions in available places in the name of sustainability all negatively affect women. Women generally earn less than men, so restrictive parking policies and high parking fees adversely affect women who are least able to pay, whereas drivers with company cars (80% male) pass the cost onto their employers. Whilst many women work in their local district, nevertheless, 80 per cent of all workers in central areas are female, including shop workers, but little would one realise this from dominant images of the 'journey to work' by the male 'commuter' (Hamilton, 2000). Many of the women working in central areas are likely to be shop assistants, clerical workers and service workers such as hairdressers and bar staff, and, from research, it would appear many endure long expensive journeys into work by public transport. Cheap housing is limited in central areas, and new accommodation generally comprises 'executive apartments' that few can afford, and that are unsuitable for family living. If women can afford a car, by economising hard, they may find that increased parking charges are beyond their means, and so they simply give up their job. Women part-time workers and those coming in 'late' because of childcare responsibilities are likely to find all spaces are already taken.

Road charging, for example with the 'Congestion Zone' in central London, is based upon the ability to pay, not upon the usefulness or purpose of the journeys undertaken. The boundaries of the congestion zone line can cut across established neighbourhoods and thus cut into women's trip training patterns, crossing in and out of the zone several times in their daily journeys. Although they only have to pay once a day for crossing into the zone as many times as they like, still they have to pay every day, for what is a substantial amount for many on low incomes, or no incomes because of caring responsibilities.

Another problem for women is the introduction of 'two plus' car lanes (where cars with only two passengers or more can travel). In their multi-purpose journeys, and trip chaining, women drivers are more likely to have passengers during some parts of their journey (children on the way to school) but not in their last lap on the way to their workplace. But lane use is based on the number of passengers present at the time that traffic camera monitors the car during one small part of what may be a complex journey. This contrasts with the more traditional journey to work, still undertaken by many male commuters, as a simple journey to and from work, with no intermediate stops, and at a fixed time each day, where, arguably, it is easier to arrange car sharing with other commuters. In contrast many women, particularly part-time and short term contract workers, and those trying to fit in a diverse range of duties into each day's journey chain are less likely to have such regular journey times and patterns. But few transport planners seem to appreciate this, being more likely to see journeys in isolation from each other. For example, a young male transport planner appeared surprised that when women are travelling accompanied by small children, by train or bus for example, they were also actively 'doing' child-care at the same time, a job more often seen as a 'static' activity that takes place within the home. An overall emphasis upon quantitative data collection that asks 'how many', rather than on a qualitative investigation that asks 'why' is extremely prevalent within transportation planning circles and particularly adversely affects women. Likewise, many main roads into cities now provide a priority lane for those with 2 or more passengers, with cameras checking the numbers of passengers. This works well for those who commute regularly to work and have workmates nearby, but it is less practical for women travelling on their own at irregular times, to dispersed locations, because local jobs are limited, and 'flexible' locum and part-time work is the only option.

Particular hatred seems to be reserved for 'the school run', that is women taking their children to school by car. This activity only contributes towards 15% of rush-hour traffic but it is widely condemned, for 'cluttering up the roads', a sentiment voiced no doubt by male commuters and businessmen who do not question their own right to use a car (Coleman, 2000). The school run is portrayed in the media as being undertaken by rich lazy housewives in their '4 by 4' Range Rovers, although many families only own a cheap car and have to make major economies to keep it running. (Seldom does one hear that some male parents undertake the school run too, that would not fit the media image.) In reality, many women find it very difficult to get to work early in the morning, to ensure their children reach the school and childminder within a very tight morning time slot. Most women's car journeys are not undertaken for their own indulgence but are the most efficient. If mothers do go by bus or train with their children during the rush-hour they may be accused of 'filling up the spaces' and slowing down the system, which is clearly not really intended for them, but for commuters (Lenclos, 2002).

Women are a sitting target for condemnation by environmentalists because in our society they are still the ones who cause pollution because they do the shopping, use washing machines, care for babies and children, and thus 'produce' household waste (and menstruate too). Another 'home-making' duty is 'putting the rubbish out', which nowadays has become a complex and heavily monitored 'recycling process', whereby 'the housewife' is expected to divide the refuse into paper, glass, plastic, tin, etc., and may be fined for sorting it wrongly. Again sustainability falls heavily on the woman, in terms of time, especially if there is not a kerb-side collection. In many local authority areas the householder is expected to take the rubbish to the official recycling bins, which involves either a car journey or taking it on the bus. On the local news there have been several reports of pensioners who have been turned off the bus for taking 'hazardous' materials on the bus, and the elderly struggling with 'wheelie bins' (which have replaced dustbins in many areas) which are far to heavy and big for the modest amounts of rubbish they generate (Oates and McDonald, 2006).

Women also feel obliged to undertake 'escort' journeys to ferry children and teenagers around because public transport is so limited, and cycling is so dangerous. Condemnation might better be levelled at local education authorities who decentralise schools, and who will not let families send siblings to the same local school, and to bus operators who refuse to see the need for a link between employment and education locations in their timetabling schedules. 'Walking-trains to school' schemes are certainly an option in higher urban density areas, but are hardly practical in suburban locations where schools are several miles away and with siblings at different schools. Such schemes are also prone to assume that some mothers will be freely available to help by volunteering early in the morning, without any worry about being late for work.

Another example of lack of 'joined-up thinking' that is essential to an integrated approach to transport policy is manifested in the continuing closure of local facilities. Throughout England and Wales, local post offices, banks, railway stations, libraries, local council offices, health centres, inter alia, are being closed down and centralised, resulting in everyone having to travel further and queue longer for essential services, particularly women who tend to undertake the main bill-paying, parcel-posting, and 'home-making' administrative duties. The 'queue', that is, standing in line, is arguably the slowest 'journey' and most frustrating form of transport there is. Talking of queues, a key factor that many women have commented upon is the importance of providing toilets, be they municipal public toilets or off street toilets in shops or cafes to which the public has right of access. 'Away from home' toilets are particularly important at transport termini. Lack of toilets limits the amount of time spent away from home and the 'bladder's leash' can be a major constraint for the elderly, pregnant and those with disabilities (Bichard et al., 2004). Public toilets are often located down steps or through turnstiles, preventing access for those with luggage, heavy shopping, and pushchairs, a limitation also encountered if trying to use the much-feared, high-tech, APC (Automatic Public Toilet) (Greed, 2003). Whilst women form the majority of public toilet users, men are the providers, designers and managers of toilets, who possess little understanding of women's toilet needs (Anthony, 2001). But, men's experience is entirely different, as they have approximately twice as many facilities, including male-only street urinals which have made a come-back in London, Amsterdam and other European cities, as culturally, 'only males are privileged to overflow on the public highway' (Chevalier, 1993, 97).

There is also a need for seats, bus shelters, good lighting, and accessible, safe, urban environments. We have an ageing population and people need the infrastructure and assistance to travel around. The decline in porter services at railway stations combined with many steps and changes of level makes travel onerous for many. Furthermore, passengers are often expected to put their heavy suitcases on a rack up above their seats with no assistance and limited alternative storage offered. Talking of luggage, the recent restrictions at airports permitting only one small piece of hand luggage seems to be particularly sexist, in that 'ladies' handbags' are not permitted, and clearly not seen as essential, whilst 'women's personal items' have been subject to restriction or degrading inspection. Again men's experience is entirely different, they put all their valuables in their jackets and are not encumbered with the need to carry emergency sanitary items.

How to bring change

This section presents one means of change: the strategy of mainstreaming gender considerations into policy-making, drawing on a Gender Mainstreaming Toolkit that was developed for the Royal Town Planning Institute (RTPI) www.rtpi.org (Reeves and Greed, 2003). Gender mainstreaming has been defined as 'the systematic integration of gender into all systems and structures [of government]. Policies, programmes, processes and projects, into ways of seeing and doing, into cultures and organisations' (Kirton and Greene, 2003). Thus gender is required to be taken into account at all stages of the plan-making process and all aspects of planning practice. Since women have a considerable amount of catching up to do, and because planning policy and practice has, empirically, been shown not to give priority to women's issues, it is likely that, as a result of operating impartial gender mainstreaming techniques, the needs of women will still be better recognised and acted upon within the planning process (Reeves, 2005).

The RTPI toolkit was developed to enable spatial planners in local authorities to mainstream gender into planning policy and practice. It primarily works at forward planning levels, where the key decisions are made that shape both urban areas and the countryside, but it can also be applied at the desktop, development control level. The toolkit is not a formula, a spatial blueprint, or prescription that has to be rigidly followed. Rather it seeks to provide a series of questions that need to be asked at each stage of the planning process, which will enable planners to judge whether gender is being taken into account in the decision-making process.

There are many questions, stages and areas of planning policy that need to be taken into account, and these are detailed in the full version (Reeves and Greed, 2003). But it can all be boiled down to the following 'mini toolkit'. This comprises a list of questions, the answers to which might form the basis for focused action and change, which will result in gender considerations being built into transport-planning policy-making (see Box 16.1).

Box 16.1 The mini version of toolkit

Checklist of questions in order to mainstream gender into the planning process:

- Who are the planners? How are they chosen?
- Which sorts of people are perceived to be the planned?
- How are statistics gathered and who do they include?
- What are the key values, priorities and objectives of the plan?
- Who is consulted and who is involved in participation?
- How are planning proposals evaluated? By whom?
- How is the policy implemented, monitored and managed?
- How is gender integrated in with other policy topics?
- Are policies monitored in respect of gender impact?

Each one of these questions can be broken down into many more in order to ensure that the gender implications of each stage of the planning process are carefully taken into account. However, every planning location is different and planners need to exercise their own professional judgement and discretion, and therefore it is up to the planning authority itself to take the principles on board and apply them as best suits them. You have to use an element of professional judgement, discretion and plain intuition to make it work for your location.

Box 16.2 Key factors to take into account

The following factors need to be considered to make the process work:

- Preliminaries: What supportive initiatives, resources and possibilities exist already?
- Planners: Who is doing the planning? Who are perceived to be the planned?
- Populations: How are statistics gathered and who do they include?
- Priorities: What are the key values and objectives of the plan?
- Participation: Who is consulted and who is involved in participation?
- Piloting: How is it evaluated? Are discussion groups used? How are they selected?
- Programme: How is the policy implemented, monitored and managed?
- Performance: Who benefits? Who loses out? What side effects are there?
- Proofing: Is gender incorporated into each policy? Does it make sense?

Conclusion: alternative approaches

Rather than blaming individual women, fundamental structural changes are required to urban form and planning policy. As an alternative to spread out, zoned, low density cities, many European women planners would like to see the 'city of everyday life', which they define as the city of short distances, mixed land uses and multiple centres as the ideal objective that would fully take into account gender considerations. Such a city structure would reduce the need to travel, create more sustainable cities that would be more accessible, whilst creating a higher quality urban environment for all. It would provide more jobs and facilities locally and help revitalise declining areas overall (Skjerven, 1993). In contrast, in some more progressive European countries, and in some parts of North America, one finds local, positive efforts to make spatial proximity links between childcare, schooling, employment and shopping functions (FCM, 2006).

Rather than introducing negative car-controlling policies first, it would be better to invest more in public transport and to use the planning control system to ensure that neighbourhoods are designed to provide local facilities, amenities and employment opportunities. Meanwhile back in Britain, it is simply not realistic to carry home a week's family shopping from the nearest, but distant supermarket, on the handlebars (or in the panniers) of a bicycle, particularly if one is working full time, is 'time-poor', and cannot even guarantee to be at home if one subscribes to a home delivery service. If the full agenda of 'sustainability' were taken into account, rather than over-emphasising the environmental dimension, then it is likely that sustainable transport policy would prioritise different types of journeys as 'essential', and more recognition would be given to the need for accessible public transport for all, whilst the routes and timetables would be reconfigured to meet the 'off peak' needs of women. A wider picture of travel patterns and their social value would be built up. Supporting services such as toilets, bus-shelters, creches, cycle lanes, steps, carriage of luggage, and shopping home delivery would all be integral components of the transportation infrastructure. Traffic control would be based upon the usefulness of the journey rather than the ability to pay.

References

Anthony, K. (2001), *Designing for Diversity: Gender, Race and Ethnicity in the Architectural Profession* (Chicago: University of Illinois).

Bichard, J., Hanson, J. and Greed, C. (2004), *Access to the Built Environment Barriers, Chains and Missing Links* (London: University College London).

Booth, C., Darke, J. and Yeandle, S. (1996), *Changing Places: Women's Lives in the City* (London: Paul Chapman).

Buckingham-Hatfield, S. (2000), *Gender and Environment* (London: Routledge).

Chevalier, G. (1993, originally 1936) *Clochemerle* (London: Mandarin).

De Graft-Johnson, A., Manley, S. and Greed, C. (2006), 'Why women leave architecture', *Construction Management and Economics* 23 (10), 1035–43.

DTI (2005), *The Mobility and Inclusion Unit Business Plan 2005-6* (London: Department of Transport in association with the Women's Transport Network).

Early, C. (2006), 'Travel choice backer', *Planning page* 13, 29 September.

FCM (2006), *A City Tailored to Women: The Role of Municipal Governments in Achieving Gender Equality*, Federation of Canadian Municipalities, Montreal, <http://www.ville.montreal.qc.ca/femmesetvilles> (available in English and French)

Gilroy, Rose and Booth, Christine (1999), 'The role of a Toolkit in mobilising women in local and regional development', presented at the Planning Research Conference, University of Sheffield, 1999.

GLA (2003a), *Local Transport Plans for London Boroughs* (London: GLA).

GLA (2003b), *Gender Equality Scheme* (London: GLA Online). Available at <http://www.london.gov.uk/gla/publications/women/gender-equal03.pdf>.

GLC (1985), *Women on the Move: Women and Transport Policy* (London: Greater London Council).

Greed, C. (1994), *Women and Planning: Creating Gendered Realities* (London: Routledge).

Greed, C. (2003), *Inclusive Urban Design: Public Toilets* (Oxford: Architectural Press).

Greed, C. (2005), 'An Investigation of the Effectiveness of Gender Mainstreaming as a Means of Integrating the Needs of Women and Men into Spatial Planning in the United Kingdom', *Progress in Planning* 64, 239–321.

Greed, C. and Reeves, D. (2005), 'Mainstreaming Equality into Strategic Spatial Policy Making: Are Town Planners Losing Sight of Gender?' *Construction Management and Economics: Special Issue on 'Diversity and Equality in Education'* 23 (10), 1059–70. Available as a pdf and through library electronic journals system.

Hamilton, Kerry (2000), *Public Transport Audit London* (London: DETR Mobility Unit undertaken by University of East London).

Kirton, Gill and Greene, Anne-Marie (2003), *The Dynamics of Managing Diversity: A Critical Approach* (Oxford: Butterworth Heinemann).

Kunstler, James (1993), *The Geography of Nowhere: The Rise and Decline of America's Man-made Landscape* (New York: Touchstone).

Lenclos, Marie (2002), *Inclusive Design: Access to London Transport* (London: Royal College of Art).

Oates, C. and McDonald, S. (2006), 'Recycling and the Domestic Division of Labour: is Green the New Pink? *Sociology* 30 (3), 417–33.

ODPM (2003), *Housing: Planning Policy Statement 3* (London: Office of Deputy Prime Minister, recently renamed, DCLG, Department of Communities and Local Government).

ODPM (2005), *Diversity and Equality in Planning: A Good Practice Guide* (London: ODPM).

ONS (2005), *Social Trends* (London: Office of National Statistics).

Reeves, D (2005), *Planning for Diversity: Policy and Planning for a World of Difference* (London: Routledge).

Reeves, D. and Greed, C. (2003), *Gender Equality and Plan Making: The Gender Mainstreaming Toolkit* (London: RTPI) and see <http://www.rtpi.org.uk>.

Skjerven, R. (1993), *Manual of Alternative Municipal Planning, Ministry of Environment* (Oslo: Ministry of Environment).

EPILOGUE

Chapter 17

Gendered Mobilities: Epilogue

Mimi Sheller

Introduction

The articles gathered in this volume are indicative of an urgent interest in the importance of gender in the planning, design, and practice of sustainable mobilities. The solution to contemporary problems of mobility will not be found without sustained attention to women and men's differential mobilities, to the gendering of design and planning processes, and to the gendered underpinnings of discourses of mobility. This is true both at the local level of homes, neighbourhoods and cities, and at the national and transnational levels of travel, migration, and globalisation.

Mobilities are never free and unfettered, but are striated by 'a whole series of rules, conventions and institutions of regulation and control ... a systematized network' (Amin and Thrift, 2002, 26). Gender distinction and inequalities are one of the crucial sets of conventions and institutions by which mobility is 'differentiated' and unevenly enacted in processes imbued with social, cultural, economic, political and geographical power (Cresswell, 2006).

This epilogue offers an overview of the various ways in which mobilities are gendered and a summing up of the contributions made by the chapters in this collection to a better understanding of gendered mobilities. I begin with gendered discourses of mobility and how they have shaped the field of mobilities research, and then consider in turn gendered geographies of mobility; gendered technologies of mobility; and conclude with some reflections on some of the more positive developments in gender research for generating alternative mobility futures.

Gendered discourses of mobility

The concepts of mobility and immobility enter into our discussions as already deeply gendered discourses. Modernity, progress, and privileged forms of masculinity have long been associated in Western thought with mobility, while immobility, stasis and sedentary states have been attributed to 'backward' societies of so-called Oriental despotism, Eastern serfdom and Medieval feudalism, for example in the works of Marx, Weber and Durkheim. The sea-faring discovery and exploration of the New World, the 'spirit of capitalism' associated with modern northwestern European Protestant cultures, and the American ideology of manifest destiny and the opening of the frontier, are just a few of the cultural touchstones that not only define the West as mobile and expansionist, but do so through masculine figures such as the explorer, the entrepreneur, and the frontiersman.

If cultural judgments about the value of mobility have historically been posed in such gendered terms, it is not surprising that tropes of home and dwelling are feminised in ways that essentialise gender difference, romanticise a home-bound femininity, and often devalue the sedentary spaces and bodies associated with women (Morley, 2000; Massey, 1994). Women are often defined as lacking a 'mobile subjectivity', being rooted in place and home, while narratives of masculine becoming often hinge on travel, hitting the road, and escape from home (Ahmed et al., 2003). Sheller and Urry (2006) thus suggest that a new mobilities paradigm must delineate the contexts in which both sedentary and nomadic accounts of the social world operate, with an understanding of how gendered discourses of mobility operate and circulate.

Because of these deep-seated gendered understandings of mobility 'nomadic theories' that celebrate metaphors of mobility, travel, and flight often rest on a romantic reading of mobility that is highly gendered (Braidotti, 1994; Cresswell, 2002, 15-18; Urry, 2000, chapter 2). Beverley Skeggs argues that the mobility paradigm more generally can be linked to a 'bourgeois masculine subjectivity' that describes itself as 'cosmopolitan' (Skeggs, 2004, 48). 'Mobility and control over mobility both reflect and reinforce power. Mobility is a resource to which not everyone has an equal relationship' (Skeggs, 2004, 49; Morley, 2000; Massey, 1993). Caren Kaplan argues that 'certain ways of seeing [arise] as a result of this privileging of cosmopolitan mobility' (Kaplan, 2006), and these ways of seeing align with what she calls 'the cosmic view' in warfare, associated with total control of airspace, which originates in a masculine, disembodied 'God's eye' view of the world. The visualising and locational technologies that support such a view are also deeply gendered in their design and applications, as I shall discuss below.

Despite these gender-differentiated discourses of mobility there is an absence of gender awareness and gender analysis in much of the policy discourse surrounding transportation, mobility and the information society. Universalistic language often masks differential processes, while gendered distinctions in the application of legal standards and moral reasoning may, for example, privilege men in the determination of acceptable approaches to traffic interaction, as Anette Jerup Jørgensen argues in Chapter 7. Furthermore, the largely male experts and technicians who dominate transport policy, planning and design may overlook women's perspectives, experiences and needs, or see them as irrelevant to the sector (see Polk, Chapter 15; Greed, Chapter 16). As Sumeeta Srinivasan argues in Chapter 10, improvements in high-speed mobility in cities such as Chengdu, China, or Chennai, India, fail to address issues of accessibility for those whose main mode of transport is walking, especially women.

When policy makers debate mobility systems, or designers implement new technologies, or researchers study new mobilities, they are unconsciously already working within a context of deeply gendered discourses that must be brought to the foreground if we are to understand how planning decisions may be contributing to unequal mobility outcomes for men and women. This book makes an important contribution to conceptualising the key questions and framing the core issues surrounding gendered mobilities, moving it as a concern to the forefront of mobilities research, and delineating and opening up some important agendas for future research.

Gendered geographies of mobility

Since time immemorial men and women have had different patterns of access to travel and movement, whether locally, nationally, or internationally, in both Western and non-Western cultures. It is a question 'not simply about *who* travels but *when, how, and under what circumstances?*' (Brah, 1996, 182, emphasis in original). We can understand this as variation not only in access to mobility, but as David Kronlid argued in Chapter 2, as a question of mobility as a 'capability' which may be more or less constrained and enabled in different historical and cultural contexts. Many cultures throughout the world have limited or circumscribed women's ability to travel and to move. Religious practices associated with female modesty (such as veiling, *purdah*, or general separation of the sexes) have prohibited women's travel or made it possible only with the accompaniment of male relatives. Cultural fashions such as foot-binding in China, corsets in the Victorian era, or high-heeled shoes in the modern era also are understood as limitations upon female mobility, according to feminist analysts, and as a fetishisation of female immobility. In more recent contexts, several chapters in this volume have pointed out that in urban areas women tend to have shorter work trips than men, and to use private automobiles less often than men.

The male body is culturally performed as a more mobile body, while the female body becomes more restricted and spatially circumscribed. Socialisation of boys and girls often produces a distinctive embodied *habitus* and 'bodily hexis', to use Bourdieu's terms, in which boys have more latitude for movement, activity, travel across space, and risk-taking, while girls tend to be enculturated into more sedentary activities, more circumscribed uses of space, and greater risk aversion (Young, 1990; and see Murray, Chapter 4 this volume; and Jørgensen, Chapter 7). Boys (at least until recently) often were accorded more freedom than girls to move around their neighborhoods, to engage in unsupervised walking or riding bicycles to meet with friends, to go camping, fishing or hiking, or to ride on public transport. Such capabilities for mobility are deeply tied up with the production of white masculinity and its materialisation in particular kinds of embodied agency. As Sheela Subramanian argued in Chapter 3, there is a 'deep link between 'race', gender and movement in space', and gendered mobilities are always also racialised. Drawing on Judith Butler and Frantz Fanon, she shows how 'bodies and spaces produce one another, and how the racialised [and gendered] body is constituted through immobility'.

Limitations on women's mobility have also been closely linked to threats of violence against women, as highlighted by the anti-violence 'Take Back the Night' campaigns of second wave feminism in the West, as well as more recently in stories of the surveillance and violent policing of women's mobility in Islamic fundamentalist societies (stories which, when told in the Western media, also help to symbolically locate 'the West' as a space in which women enjoy greater freedom of mobility). Freedom of mobility, however, is not only unevenly distributed in Western societies according to gender, age, and race, but also may in some cases occur as a kind of constraint. As Lesley Murray argued in Chapter 4, in societies based on automobility the increasing expert-management of risk combined with expectations of certain forms of 'good mothering' constrain (middle-class) mothers into highly flexible forms of 'hyper-mobility', as they feel the need to escort their children everywhere. Automobility and other forms of personal mobility may be experienced as liberation, yet at the same time may produce various kinds of constraints and coercions (Sheller and Urry, 2000; Urry, 2000; Jørgensen, Chapter Seven), for both men and women.

Women's mobility has long been linked to concerns over prostitution and other forms of sexual-economic transaction that take place outside the home (Kempadoo, 2004). In recent times this set of concerns has been caught up in movements against the 'trafficking in women', in which men are alleged to be the 'traffickers' of women; but international sex work often involves complex intersections between migration, mass media, tourism, and diaspora. As Arjun Appadurai notes in reference to Keralan women entertaining men in Bombay clubs, or Japanese and German tourists engaging in sex tourism in Bangkok, deterritorialised money, commodities and persons intersect with the 'mediascapes' of film, entertainment, and pornography to produce 'loops which tie together fantasies about the Other, the conveniences and seductions of travel, the economics of global trade and the brutal mobility fantasies that dominate gender politics in many parts of Asia and the world at large' (Appadurai, 1996, 39). Tourist fantasies and seductions of travel just as strongly influence Western women's travelling sexualities, as Susan Frohlick suggests in Chapter 9, raising questions about 'mobile sexualities' and the construction of racialised bodies in 'hot' locations such as the Caribbean. Women may experience the mobilities associated with migration and with tourism as simultaneously liberatory and coercive, enabling new projects of the self, yet constraining them into certain gendered/racialised performances. Nadine Cattan (Chapter 6) also shows how researchers need to attend to the more polycentric and flexible configurations of women's migratory patterns in order to generate more complex (and realistic) geographies of mobility.

Both intra-regional and international migrations are deeply gender-differentiated processes, both in terms of the kinds of 'labour force' that are mobilised by international capital and in terms of the mobility fantasies that feed migrant dreams and projects. Migration chains often begin with either male (day labour, construction, farm work, restaurant work) or female (factory work, chamber maids, nurses, care-givers) economic opportunities, leading to skewed migration patterns from particular communities and within communities of arrival. Families may eventually follow the pioneering migrant, but the uneven economic opportunities associated with migration often have long-term affects on the gender dynamics within such families, including access to financial resources, commodities, and new media, 'ideoscapes' and identities. Again Appadurai has pointed to the significance of gendered mobilities in the reproduction of culture, 'Women in particular bear the brunt of this sort of friction, for they become pawns in the heritage politics of the household, and are often subject to the abuse and violence of men who are themselves torn about the relation between heritage and opportunity in shifting spatial and political formations' (Appadurai, 1996, 44).

Family politics and sexual politics are at stake as people, commodities and capital move around the world. As Elisabeth Scheibelhofer found in Chapter 8, even more privileged groups of highly mobile professional women from Western Europe find themselves responsible for 'the gendered care for social relations', such as caring for elderly parents, which over the life-course restricts their mobility; 'gender gives shape to mobility and migration' amongst all socioeconomic groups. Migration is also associated with global networks of sending monetary remittances to home countries, and these may also have significant gender differentials in terms of their control and distribution at the receiving end, impacting even on those who have not moved. Domestic politics in the sense of both the politics of the home and the politics of the nation can no longer be separated from these complex transnational implications of gendered mobilities.

Gendered technologies of mobility

Mobilities are highly dependent on technological innovations, and technology is itself highly gendered in its design, distribution and systemic affects (Wajcman, 1991). When new technologies of transportation are invented, they are usually expensive and the first adopters tend to be elite privileged men (think of horses, personal carriages, bicycles, automobiles, and airplanes, each of which was first used mainly by wealthy men). In some periods and some cultures women have been prohibited from using some of these technologies, as in Western societies where bicycles were initially not designed for women, in the early civilian use of airplanes mainly for male businessmen, and in contemporary Saudi Arabia where women cannot drive cars. Patriarchal beliefs about the inappropriateness of women's spatial mobility outside the home continue to influence research and planning in the field of transport. Even where 'gender mainstreaming' has been introduced into government transport policy, it has not been easily translated into successful action and concrete strategies (see Polk, Chapter 15; Greed, Chapter 16).

Several chapters in this volume consider women and men's differential access to automobility. Petter Næss (Chapter 12) shows how gender differences in urban daily-life travel in Copenhagen are partly derived 'from differences in activity patterns and responsibilities' yet are also in part due to gendered differences in access and attitudes toward different modes of travel, wherein it was often assumed that men had priority over women in familial car use. And Randi Hjorthol's (Chapter 13) research on Norway suggests that women's greater responsibility for household tasks limits their daily travel, and that men have greater access to cars and use them for more trips for themselves, rather than for others. However, Nadine Cattan suggests in Chapter 6 that women's home to workplace mobility may actually depend more on gendered spatial structuring of job markets than on personal choices or 'familial' decisions. Either way, ungrounded beliefs and assumptions about women's commitment to childcare and familial responsibility still influence transport planning and determine men's greater mobility capability.

There is also a longstanding relation between technologies of mobility and a masculinist military-industrial complex. European patterns of road-building, for example, date back to the building of Roman roads for the movement of troops in straight lines across conquered lands, or the clearance of convoluted urban slums for the building of boulevards accessible to police and militias, as in Haussman's restructuring of Paris. Air transport is also an offshoot of military research and development, first of early prop-engines and later helicopters, jet engines, rockets and recent stealth technologies. Each of these has supported largely male militaries and even when they made the leap to civilian uses remained largely in the hands of men initially. Mobile communication technologies are also dependent on military research and development, as in the origins of radio, the Internet, global positioning systems (GPS) and geographical information systems (GIS) as military technologies that were transferred over to civilian uses (Kaplan, 2006).

Mobility is not simply locational, but is also imaginative, ideational, and linked to the 'virtual' technologies of media in the Internet age. Recent developments in mobile technology are leading to changes in the spatial and temporal ordering of social practices for some social groups, which are just beginning to be researched (for example, Brown, Green and Harper, 2002; Urry, 2002; Wittel, 2001). It is indisputable, however, that men constitute an overwhelming proportion of technical disciplines and high-tech developers, thus as with earlier technologies there is likely to be an in-built male-bias in their design and marketing (see Inkinen, Chapter 14). More broadly speaking, the 'mass diffusion of information technologies and automobiles, along with the simultaneous production and organisation … of multiple and generic built spaces that are intimately coordinated across vast distances', argue Graham and Marvin, 'are about the architectural shaping of time as well as spaces' (Graham and Marvin, 2001, 30). The convergence and blurring of spaces and times of business, leisure, travel, and dwelling for certain groups (that is, especially young professionals) is said to be producing novel pressures to manage fragmented time budgets and dispersed social contacts in more complex ways (Wittel, 2001; Sheller and Urry, 2006). Because of uneven access to existing public space and to new transport and communication technologies, men and women face different problems of coordination and innovation in the use of new mobile technologies.

Research on mobile communication technologies and the Internet tends to show differential patterns of usage amongst men and women, as Inkinen (Chapter 14) points out, and also indicates that gender differences in technology usage tend to be overlooked in information society strategies. Thus it may be important to put gender on the agenda in the discussion of mobile communications and the relationship between physical mobilities and internet-based networking. As Michaela Fay suggests in Chapter 5, we must pay attention not only to physical mobility but also to the imagining of mobility, in particular 'the imagining of a network of mobile feminists' who have a complex engagement with mobility and community-building. As women engage with online community-building, transnational networks, and new forms of political and affective attachments, they may be inventing new kinds of gendered mobilities.

We might also ask what impact these changes in social time-space are having on the gendered structures of public and private life, public and private space, and the public and private spheres. Without assuming that everyone takes part in such social practices (given the unevenness of new technology adoption, the social stratification of cultural practices, and the powerful forces of social exclusion), we can nevertheless see that the new temporalities and spatialities associated with these complex mobilities may have far reaching implications for the entire fabric of spaces of public and private life, with particular relevance for the gendered structures of inequality that have long informed the public/private distinction (Fraser, 1992). Further research is needed into how new technologies of mobile communication are changing the ways in which men and women navigate, perform and perhaps transform the public/private distinction in newly gendered ways.

The new kinds of 'personalised scheduling' afforded by mobile devices ineluctably come with systems of sorting, tiering, channelling, and differentiating access for 'preferred customers' (see Sheller and Urry, 2006). Third generation wireless communication systems in the United States, for example, claim to offer 'ubiquitous' and 'instantaneous' communications for business customers moving across a world of seamless high-speed connections enabled by expensive new devices, but such premium services are predicated on the dismantling of the ideals and infrastructures of universal access that once underpinned the public utilities and the social inclusion of the entire 'public' in urban public space (Graham and Marvin, 2001). The decline of public infrastructures and universal access has particularly hard impacts on women, who are less likely than men to be the 'preferred customers' of business elitism. The uneven channelling of 'First World' and 'Third World' levels of service, and the splintering of access within urban polities, and the creation of enclaves or corridors of high connectivity, will also have significant gender effects.

Conclusion

In conclusion, this volume has presented a wide range of research on interlocking topics in the construction of gendered mobilities, including issues of mobility capability and access, automobility and transport planning, travel and migration, mobile communications and use of the Internet. The focus in many chapters has been on women's lack of full access to transportation and other technologies, exclusion from technical design and governance of systems of mobility, and continuing inequality in mobility capability at many different levels. Yet the positive message here is that much research is being undertaken in these areas, new models are being generated, and we seem to be on the cusp of better understanding of the causes behind the production of uneven mobilities.

Even more optimistically, we may hope that emerging processes of redesigning the technologies, spaces, and practices of mobility may be especially beneficial to women. For example, trends toward the design of more flexible systems that call for more 'soft' vehicles, multi-modal stations, and integrated urban transport systems may benefit women as much as, or even more than, men. Greater awareness of the incompleteness of gender mainstreaming and ongoing issues of social exclusion may help to promote better research into underlying causes and comparative analysis of best practice. Growing reflexivity around the gendered dimensions of mobility discourses may help women to more effectively intervene in debates around issues affecting mobility, whether in relation to transport, tourism, travel, or migration.

Indeed, the integration of these sub-fields into a single field of mobilities research may help to generate new perspectives and approaches to longstanding problems.

Gendered Mobilities addresses a complex range of issues spanning First World and Third world locations, qualitative and quantitative methodologies, theoretical questions and more pragmatic policy questions, and scientific as well as normative approaches. This wide-ranging agenda delineates the parameters for future research and lays out markers for both academics and policy-makers to pay attention to in all future planning and research on mobilities.

References

Ahmed, S., Castaneda, C., Fortier, A. and Sheller, M. (eds) (2003), *Uprootings/Regroundings: Questions of Home and Migration* (Oxford: Berg).

Amin, A. and Thrift, N. (2002), *Cities: Reimagining the Urban* (Cambridge: Polity Press).

Appadurai, A. (1996), *Modernity at Large: Cultural Dimensions of Globalization* (Minneapolis and London: University of Minnesota Press).

Brah, A. (1996), *Cartographies of Diaspora: Contesting Identities* (New York and London: Routledge).

Braidotti, R. (1994), *Nomadic Subjects: Embodiment and Sexual Difference in Contemporary Feminist Theory* (New York: Columbia University Press).

Brown, B., Green, N. and Harper, R. (eds) (2002), *Wireless World* (London: Springer).

Cresswell, T. (2002), 'The Production of Mobilities', *New Formations* 43, 11–25.

Cresswell, T. (2006), *On the Move* (London: Routledge).

Fraser, Nancy (1992), 'Rethinking the Public Sphere', in C. Calhoun (ed.), *Habermas and the Public Sphere* (Cambridge and London: MIT Press), pp. 109–42.

Graham, Stephen and Marvin, Simon (2001), *Splintering Urbanism: Networked Infrastructures, Technological Mobilities and the Urban Condition* (London and New York: Routledge).

Hannam, K., Sheller, M. and Urry, J. (2006), 'Editorial: Mobilities, Immobilities and Moorings', *Mobilities* 1 (1), 1–22.

Kaplan, Caren (2006), 'Mobility at War: The Cosmic View of US 'Air Power'', *Environment and Planning A* 38 (2), 395–407.

Kempadoo, K. (2004), *Sexing the Caribbean: Gender, Race and Sexual Labor* (London and New York: Routledge).

Massey, D. (1993), 'Power Geometry and a Progressive Sense of Place', in J. Bird et al. (eds), *Mapping the Futures: Local Cultures, Global Change* (London: Routledge), pp. 59–69.

Massey, D. (1994), *Space, Place and Gender* (Cambridge: Polity Press).

Morley, D. (2000), *Home Territories: Media, Mobility and Identity* (London: Routledge).

Sheller, M. (2003), 'Mobile Publics: Beyond the Network Perspective', *Environment and Planning D: Society and Space* 22 (1), 39–52.

Sheller, M. and Urry, J. (2000), 'The City and the Car', *International Journal of Urban and Regional Research* 24 (4), 737–57.

Sheller, M. and Urry, J. (2002), 'Mobile Transformations of 'Public' and 'Private' Life', *Theory, Culture and Society* 20 (3), 107–125.

Sheller, M. and Urry, J. (2006), 'The New Mobilities Paradigm', *Environment and Planning A* 38, 207–26.

Skeggs, B. (2004), *Class, Self, Culture* (London: Routledge).

Urry, J. (2000), *Sociology Beyond Societies* (London: Routledge).

Urry, J. (2002), 'Mobility and Proximity', *Sociology* 36, 255–74.

Wajcman, J. (1991), *Feminism Confronts Technology* (Cambridge: Polity).

Wittel, A. (2001), 'Towards a network sociality', *Theory, Culture and Society* 18, 31–50.

Young, I.M. (1990), *Throwing Like a Girl and Other Essays in Feminist Philosophy and Social Theory* (Bloomington: Indiana University Press).

Index